CAMBRI

Travel and Exploration

The history of travel writing dates back to the Bible, Caesar, the Vikings and the Crusaders, and its many themes include war, trade, science and recreation. Explorers from Columbus to Cook charted lands not previously visited by Western travellers, and were followed by merchants, missionaries, and colonists, who wrote accounts of their experiences. The development of steam power in the nineteenth century provided opportunities for increasing numbers of 'ordinary' people to travel further, more economically, and more safely, and resulted in great enthusiasm for travel writing among the reading public. Works included in this series range from first-hand descriptions of previously unrecorded places, to literary accounts of the strange habits of foreigners, to examples of the burgeoning numbers of guidebooks produced to satisfy the needs of a new kind of traveller - the tourist.

Memoirs of Lieutenant Joseph René Bellot, With His Journal of a Voyage in the Polar Seas in Search of Sir John Franklin

Joseph René Bellot (1826–53) was a French naval officer whose travels took him from Africa to the Arctic before his tragic death at the age of 27. In 1851 he joined a British expedition to search for the missing polar explorer Sir John Franklin (1786–1847), whose expedition to find the North-West Passage was last heard of in July 1845. Although the voyage was unsuccessful in its search, it explored previously unknown areas of the Arctic. Bellot kept extensive notes about his journey in this remote region; they originally appeared in French in 1854 and were translated into English in 1855 and published in two volumes. In Volume 2, Bellot, who was regarded as a hero in both France and Britain, describes how the crew survived the harsh climate of the Arctic winter, his exploration by dog-sledge of inland polar regions, and his eventual return to Britain.

Cambridge University Press has long been a pioneer in the reissuing of out-of-print titles from its own backlist, producing digital reprints of books that are still sought after by scholars and students but could not be reprinted economically using traditional technology. The Cambridge Library Collection extends this activity to a wider range of books which are still of importance to researchers and professionals, either for the source material they contain, or as landmarks in the history of their academic discipline.

Drawing from the world-renowned collections in the Cambridge University Library and other partner libraries, and guided by the advice of experts in each subject area, Cambridge University Press is using state-of-the-art scanning machines in its own Printing House to capture the content of each book selected for inclusion. The files are processed to give a consistently clear, crisp image, and the books finished to the high quality standard for which the Press is recognised around the world. The latest print-on-demand technology ensures that the books will remain available indefinitely, and that orders for single or multiple copies can quickly be supplied.

The Cambridge Library Collection brings back to life books of enduring scholarly value (including out-of-copyright works originally issued by other publishers) across a wide range of disciplines in the humanities and social sciences and in science and technology.

Memoirs of Lieutenant Joseph René Bellot

With His Journal of a Voyage in the Polar Seas in Search of Sir John Franklin

Volume 2

Joseph René Bellot

CAMBRIDGE UNIVERSITY PRESS

Cambridge, New York, Melbourne, Madrid, Cape Town,
Singapore, São Paolo, Delhi, Mexico City

Published in the United States of America by Cambridge University Press, New York

www.cambridge.org
Information on this title: www.cambridge.org/9781108050050

© in this compilation Cambridge University Press 2012

This edition first published 1855
This digitally printed version 2012

ISBN 978-1-108-05005-0 Paperback

This book reproduces the text of the original edition. The content and language reflect
the beliefs, practices and terminology of their time, and have not been updated.

Cambridge University Press wishes to make clear that the book, unless originally published
by Cambridge, is not being republished by, in association or collaboration with, or
with the endorsement or approval of, the original publisher or its successors in title.

MEMOIRS

OF

LIEUTENANT
JOSEPH RENÉ BELLOT,

CHEVALIER OF THE LEGION OF HONOUR,
MEMBER OF THE GEOGRAPHICAL SOCIETIES OF
LONDON AND PARIS, ETC.

WITH

HIS JOURNAL OF A VOYAGE IN THE POLAR SEAS
IN SEARCH OF SIR JOHN FRANKLIN.

IN TWO VOLUMES.

HURST AND BLACKETT, PUBLISHERS,
SUCCESSORS TO HENRY COLBURN,
13, GREAT MARLBOROUGH STREET.
1855.

The English Copyright of this Work is the property of the Publishers.

J. CLAYTON, PRINTER, 10, CRANE COURT, FLEET STREET.

JOURNAL OF A VOYAGE

TO THE

POLAR SEAS.

CHAPTER I.

EXCURSION TO PORT LEOPOLD.

26th October.—HURRAH! I have returned, and am at last happy, having brought Mr. Kennedy and his four men back in perfect health. On Wednesday, the 15th, notwithstanding all the difficulties of equipping our men, it was decided after breakfast that eight men should go with me to the rescue of our property; as we were to pass the night out, they all wanted some additional clothing for their comfort. At ten o'clock we set out, dragging our *you-you* over the smooth and slippery ice, and by four o'clock we had reached the spot of our former disaster. After some search, we discovered our baggage, borne away by the current some hundred *mètres* more to the north, in the midst of icebergs, grinding against one another. After steering our boat through several pools of water lying over the hummocks, we found our goods frozen up in the ice, and had to break through it to get at them. After labouring for two hours, we were fortunate enough to recover everything, except a spade and our *conjuror* (portable kitchen). The tent had been pitched, and after supper I imparted to my company the plan I had formed.

The sledge was broken; to return on board, repair it, make preparations for a third voyage, and return to the place we were then in, would take us a week; the weather, which had been hitherto fine, would change before long; the rapidly increasing shortness of the days would augment the difficulties; and I finally proposed to them that we should, all together, continue the journey as far as Port Leopold. The provisions for four men, which we had saved, might suffice for nine; besides, I expected to find a small deposit of provisions, left by a party of travellers, some eight or ten miles south of Cape Seppings. Though we were ill provided with clothing, having come with the intention of spending only one night and day out, we should find abundance at Port Leopold. I announced beforehand that I should go, if I only found four men willing, but that a larger number would increase our chances of success.

There were no objections, except to foresee difficulties which we could surmount; as to such as we must expect to meet with, I was decided to dare them with the brave fellows who would be willing to accompany me. I gave them ten minutes to reflect; and, having consulted Mr. Anderson, the third ship's officer, Mr. Grate, the boatswain, and the rest, I saw with pleasure that all acceded to my plan. I was, however, obliged to send one man back to inform Captain Leask of my determination, and relieve him from uneasiness. I told him that I perfectly understood the responsibility of this step, and that my prudence must be all the greater, because he remained on board with only the doctor, Mr. Hepburn, and one other man; not that the ship needed its crew at the present moment; but, if any accident happened to us, it is very evident that the safety of those remaining behind would be seriously endangered.

Each man was anxious not to be the one sent back; but I settled the difficulty by choosing the one amongst us who was worst equipped. I greatly regretted that the doctor was not with us, both for the need of his services, which I expected, and because I knew how much he wished to join our expedition; but other considerations were of still greater importance.

At daybreak we set out in a northerly direction, drawing our sledge along with some difficulty over ice covered with melted snow; we walked along the coast whenever the ice was too much broken, and whenever we found a kind of shore. At all these points the floe was detached from the land, and the ice impractible; we then had to wade through snow, into which we sank up to our knees. It now struck us that the ice was no longer in the state of floe; and at a headland, six miles north of our encampment, the pack, quite close to us, and driven by a gentle southerly breeze, was going northwards at the rate of four or five miles an hour.

After reflecting on what was best to be done, I told my companions that my plan now was, to reach Elwin Bay, either with or without the aid of the boat; to leave our tent and all our baggage there; and then, with the boat alone, and no provisions of any kind, in order to be more free, I would take with me three men, and reach Port Leopold in one day, feeling certain that we could make a path for ourselves, either by land or over the pieces of ice which we should find. If we did not succeed in one day, a hole in the snow, such as the Esquimaux make, would serve as a shelter for the night; God would provide the rest. We had, however, the satisfaction of finding that a slight belt of ice ran all along the coast—not always a very wide one; but we were not disposed to be exacting.

Rendered cautious by the previous accident, I caused the second officer, Mr. Anderson, to keep fifty *mètres* in advance, sounding the ice with a boat-hook. However, notwithstanding all my precautions, the boatswain fell through a crack, which he might have avoided; but every one knows how difficult it is to make sailors prudent. We were able to change his clothes immediately, by each contributing a portion of our own dress; but we did not desire the repetition of such a bath, for we were so ill-provided, that the first had nearly exhausted our resources. At four o'clock we reached the northern shore of Elwin Bay, and at six we encamped four miles further to the north, in a sheltered ravine, where we were enabled to pitch our tent and make a fire. I rejoiced in the sleep obtained by my men; but, for my own part, I passed a sufficiently restless night, being uneasy at the character assumed by our enterprise, which I had hoped to render decisive. I could not forbear smiling at the numerous contrasts which distinguish my present life. I am many thousand miles distant from my country, commanding men of a foreign nation; an officer of a military-marine service, I am among men bound solely by a civil engagement; a Catholic,

I endeavour to keep alive in their minds a different religion in which they have been educated, and the precepts of which I deliver to them in a tongue which is not my own; nevertherless, I cannot complain of these circumstances, so widely different from the original conditions of my life; for there is not one of these men who does not regard me as a countryman, and obey me as if I were really so. The reason is, that we are all united by a common principle, and all our actions have a common aim.

If such mutual understanding be possible among individuals, to the annihilation of all differences of origin, race, religion, and language, what is there to prevent nations from forming a similar union of efforts directed towards a common end? Noble and sublime centralisation of heart and intellect, co-operating in the amelioration of the creature, for the glorification of the Creator!

Refreshed by an interval of untroubled repose, we gratefully christened the ravine by the name of *Rescue-Ravine*, and we resumed our march after a short prayer. At the distance of thirty paces from our tent, we found on the snow which had fallen during the night the tracks of a bear, which had no doubt scented our encampment, but the sonorous snoring of some of us had probably scared him, for he had not even ventured to approach a small *cache* of pemmican which I had concealed at the entrance of the ravine in readiness for our return.

Notwithstanding the repeated assurances of my Hudson's Bay travellers, who all laughed at the idea of such precautions being necessary, I resolved that in future we would always have our arms with us in the tent, and close at hand. Not that I consider it necessary, as some persons recommend, to establish a watch to prevent the approach of these animals; an officer should order nothing that it is impossible to perform; and I am convinced that when a man has walked for twelve or fourteen hours through snow, no order to remain awake— not even the dread of being visited by a wolf or a bear—would have power to banish sleep. The ice having become less hard, and being covered with frozen snow, made it very hard work to haul the boat, and our difficulties were soon augmented by snow. In spite of the most careful observation, we were unable to discover the dépôt of provisions on which we had reckoned, and which was probably either covered by the preceding winter's snow, or devoured by foxes. However, our interest in the search diminished in proportion as did our distance from Port Leopold. On every side of us arose high cliffs, ascending either perpendicularly, or in terraces like the steps of an amphitheatre; this inhospitable coast presenting no shore, no beach which would form a road in summer. We had travelled sixteen miles the first day—twenty-four from Point Wreck to Rescue-Ravine—and we imagined we had still about twenty miles to accomplish; but the appearance of these high lands is very deceptive; they seem to fly before you, never diminishing in distance; and this illusion greatly increases the moral fatigue of the journey. The sky of a gloomy slate-colour; the running water at some *mètres* distance from us, of a dull green, contrasting forcibly with the dead whiteness of the icicles; the snow which fell on our heads and crackled beneath our feet; all concurred to fill our hearts with evil presentiments. Besides, as we returned towards the north, we recognised the scenes of former sorrows: here we had beheld the last rocket; at the foot of this ravine the boat had been seen for the last time; and the painful emotions which we had undergone on the 9th of September revived in us at sight of these snow-covered cliffs. A bear, awakened by the noise made by our small party, distracted our attention, fortunately, from these sad recollections; but our numbers alarmed him, and before the gun and two pistols, which composed our arsenal, could be drawn from their cases, he had swam across a little lake, which rendered him safe from our pursuit. Our attention was also attracted for some moments by a duck, which the refraction of the atmosphere made us take at first for a large seal. Each prominent rock—every object of unusual form—was carefully observed by us; for we feared to discover in it the remains of the boat which Mr. Kennedy had had with him. About three o'clock we reached Cape Seppings, and saw before us

Point Whaler, with the tent erected for Sir John Franklin, now the only shelter which could have received our friends.

We discharged our arms repeatedly, at short intervals, in the hope that the echoes of the bay might convey these sounds to their encampment, and announce to them their approaching deliverance: on our side we listened anxiously for any sound in answer. The increasing snow concealed the land before us from our sight.

Cape Seppings is about three or four miles distant from Point Whaler. We anxiously examined the ice in search of some indications of the neighbourhood of man; but the earth was as silent as the air had been: we ceased to converse together, and the monotonous echo of our wearied steps alone broke the solitude. I had intended to avail myself of my privilege, and run on in front; but I now gladly sought a pretext for remaining in the duty of assisting my men in their difficult task of hauling the boat. A mile from the tent the sky cleared a little, and with my glass we could discover a black mass close to the sloop; but was it a boat or the steam-engine? At last we imagined we saw moving objects; I could resist no longer, and set off at full speed, promising my companions that I would speedily bring them certainty. Some minutes afterwards my hurrahs announced to them that our friends were in front; but as I approached I perceived they were but four in number, and that discovery damped my joy. Who, then, was absent? They on their side advanced rapidly, and soon we embraced each other with all the joy of friends who had never expected to meet again; for their anxiety about our fate had been equal to that which we had suffered for them. Their fifth companion had remained behind, they having entertained doubts as to the exact nature of the noise they heard when the discharge of our fire-arms first arrested their attention.

A hasty inspection of our friends reassured us as to the state of their health; their flourishing appearance bore witness that they had fully profited by the provisions deposited at Port Leopold, and we laughed heartily at their long beards and singular accoutrements. The temperature not admitting of prolonged congratulations in the open air, Mr. Kennedy did the honours of this residence, which Snow so appropriately named last year the Camp of Refuge. Finding the tent left by Sir James Ross torn in several places, and too large to be thoroughly warmed by five inhabitants, they chose the boat as more convenient; and, having covered it over with the remains of the tent and some sails, they divided it into several compartments, all tolerably comfortable, even to us, who had so recently left our more solid habitation. A stove placed on the platform of the steam-engine warmed the dining-room, which served likewise as kitchen and dormitory; a partition divided them from the vast heap of provisions sent out for Sir John Franklin and his men by the English Government.

Intoxicated with joy, we did due honour to the hospitaility of our Leopolders; and, having satisfied our appetites, were soon engaged in that most delightful of all conversation — the mutual recounting of our anxiety and sufferings. But why talk of our sorrows? Were they not all forgotten in this moment of happiness? Why remember those painful moments, now that we were once more united? Delighted to know that the ship was safe in a bay situated so conveniently within reach of the ground of our operations, Mr. Kennedy told me that, when separated by the ice from the *Prince Albert*, he was very much surprised not to see it in the morning, for we had kept so close in shore, that the projections of the coast hid us from them. He supposed that we had either been carried away south, or returned to Port Bowen, and perhaps been driven out of Lancaster Sound, as Sir James Ross was, and obliged to return to England.

Making up his mind at once with coolness and decision, and ably seconded by the brave men who accompanied him, he resigned himself to taking up his winter quarters at Whaler Point, with the intention of seeking for us during the winter, as well as continuing the search for Sir John Franklin. They were about to start, as I feared they would, first for Port Bowen, and then for Fury Coast: we had fortunately arrived in time to save them from the perils which they would have run in such an undertaking. Night was far advanced before sleep put an end to our conversation; and I several times woke up thinking I must have dreamed, and had to look carefully round me to make sure that it was a reality. Oh! what joy, if we are destined to experience the same emotions on Sir John Franklin's account! One of the things which gave me most pleasure was to see that Mr. Kennedy had always reckoned on me, and that the men who accompanied him shared his confidence in the efforts which they all felt I should not fail to make. Although our men were not much fatigued, it was impossible to think of work, and the smallest details, the minutest incidents of our respective adventures, were greedily listened to. Our friends had felt but one want unsatisfied: there were a few newspapers at Whaler Point, but no books, and to Mr. Kennedy especially, the absence of a Bible was the greatest privation. The care of the Government had provided for every other possible want. The wild animals had broken into a few provision barrels; but the remainder, though scattered by the winds and the ice, was in good condition.

No thanks were ever offered up to God with greater sincerity than ours: our hearts were even more full of gratitude than our lips could express. One of our first questions had been, if there was any news either of the *Erebus* and *Terror*, or of Commodore Austin's squadron, or of any other of the ships sailing like ourselves in the Arctic Sea. There is no news since last year, either because the ships were kept all the summer in the west, or because they have returned to England for want of power to enter Port Leopold, being blocked out by the ice, as we were at our first attempt. Mr. Kennedy would have liked to send a party to Griffith Island, to try and communicate with the squadron; but the state of the ice made that impracticable. The anxiety on board our ship would have been so great, that we must have sent them word of our fortunate meeting. My conviction, and Mr. Kennedy's too, is, that at this somewhat late period of the year, none but large parties can travel with any security. Our travelling preparations were not extensive; yet it took two days to construct a sledge for the boat: it was also necessary to make a fresh pair of shoes out of linen for all my men, who had, like myself, come with only their usual clothing. We took advantage of a few moments of leisure to take a survey of the bay of Port Leopold; it is formed by an important cape, Cape Clarence, which stretches, first to the east, then to the south, and is connected with the land to the west by means of a comparatively low neck of land. The bay is situated at the meeting of the openings of Barrow Strait, Lancaster Sound, the Wellington Canal, and Regent's Inlet: the four winds seem to have chosen it for their dwelling-place; and the north wind especially, forcing its way into the sort of funnel which forms the head of the bay, always blows furiously in there, however slight it may be elsewhere. The ruins of five or six Esquimaux huts had been found on the east coast of the bay, together with some whale's bones; among the latter was a piece resembling ivory, which made a capital

lining for the cross pieces of our sledge.
Towards the head of the bay, six graves had
been left by the *Enterprise* and *Investigator*
during their stay in the winter of 1848-49;
amongst others, that of Mr. Mathias, one of
the surgeons to the expedition. The good
feeling and piety innate among seamen never
fail to show themselves on such occasions;
and I have always seen them guard with
religious care those remains which remind
them of the uncertainty of all human existence, and of the chances to which they are
themselves exposed. The state of these
burial-places, moreover, indicated proper attention on the part of those whom God had
spared. A few simple, heartfelt lines were
inscribed on one of them; and we did not
recover for some time from the impression
which the idea of death, in a distant land,
always makes upon one! Does not our instinctive repulsion at the idea that our remains must rest far away from all that we
hold dear, recal us to the sublime belief in
the immortality of the soul?

During the whole of our stay, the wind
had blown very keenly from the north and

from the east, and the snow had fallen very thickly; so that we expected to meet with much greater obstacles than at our first voyage: but Mr. Kennedy suggested putting up a large sail in the boat as it stood in the sledge; and, as the wind was blowing at our backs, this was of so much service, that when we started on the morning of Wednesday, the 22nd, we had to run after our baggage as it glided along before us. Three cheers expressed our gratitude to the Camp of Refuge for its welcome hospitality; and we joyfully set forth at a running pace, on that very road which we had followed five days before, weighed down by anxiety and terror. Night overtook us before we found a fitting shelter, and we were ogblied to encamp on the ice, in a kind of *cul de sac* formed by two large icebergs. We were so blinded by the wind and the snow, that all thirteen of us were compelled to spend the night in a tent, nine feet long and six wide: having brought into it as much of our baggage as we possibly could, we sat down, and each contributing his share, we endeavoured to smother our sense of discomfort with songs and laughter. If any belated bear passed near us, he must certainly have been alarmed by these loud and confused sounds.

There is nothing like a bad night for travellers anxious to get on. Long before dawn we were ready to start. Our boat, which did not admit of an easy arrangement of our goods, was left behind until the spring, and we continued our road, which the state of the ice, broken by the still raging storm, rendered more and more painful and difficult. Notwithstanding the peaceable intentions with which we were all animated, never did any party look so like a band of robbers. The small icicles formed in our long beards did not serve as an ornament to us. Setting about it rather sooner than on the preceding day, we chose our ground in time to build a snow-house large enough for five of us, and a refreshing sleep recompensed us for the previous uncomfortable night. The third day, we camped a few miles to the north of Point Wreck, in the same order, feeling our vigour and energies reanimated in consequence of custom, practice, and our near approach to our destination. The weather had cleared up, and we enjoyed at our ease the picturesque effect of our little encampment. The transparent snow of which our house was formed, emitted greenish gleams of light, which gave it the appearance of a fanciful illumination; a few paces off, our canvas tent, whence a sonorous and infecting laugh was continually heard; still further, a wall of snow sheltering our modest kitchen, which consisted of a cauldron to melt the snow: in fact, as we travelled like gentlemen for whom nothing was too good, we indulged in the luxury of a cup of tea to wash down our pemmican. Round this fire, a few weaklings striving to warm themselves, bringing bits of wood carefully laid aside, another blowing the half-extinguished charcoal to light a rebellious pipe: a true scene of the Russian campaign; it was like a picture, to see these hungry men in snow-covered garments scattered over this lovely white sheet, and striking their feet against the ground in order to warm them. The bustle caused by this small number of human beings soon subsided: all was tranquil as before, and, thanks to the day's fatigues, we were soon buried in a sleep as sound as if we had been wrapped in the softest eiderdown. The snow melted in sea-water does not congeal as quickly as in fresh water; and on Saturday, the 25th, as well as on the two preceding days, we had to drag our sledge through half a foot of melted snow, which froze our feet whenever we stopped for a moment; the cold was endurable, however, as long as we kept in exercise. About eight miles from the ship the ice no longer offered us a safe pathway; and, as night was advancing, it became imprudent to venture on an unknown track in the dark; there was no favourable spot for an encampment; we were forced to leave our sledge, and hasten to make a path for ourselves over the snow at the foot of the coast. Fortunately, the ice in the bay was whole, and, towards six o'clock, Captain Leask and the others on board rejoiced, in their turn, at a meeting of which they had almost despaired. Mr. Leask, without believing in so complete a success, had approved of the determination taken on the 15th, as the only plan practicable and likely to succeed; and he was not uneasy at our absence having been longer than I announced, thinking that the unfavourable weather must have delayed us. Personally, I rejoice the more at having adopted this course, because I see plainly that we could not begin now with the same chances of success as we had then; and in that case we should, perhaps, have been compelled to wait until next spring.

I observed a curious phosphorescence in the melting ice on Saturday morning, the 25th, and have seen nothing like it in any other place where the snow is mixed with salt water. Is it caused by the presence of a foreign body—the particles of a fish, for instance? Whatever be its cause, the effect is this: the sledge left behind it long fiery tracks, and our footsteps seemed to strike out sparks all the time we passed along the same piece of ice.

27th October.—To-day I went with twelve men and our dogs in search of the sledge we left behind us on Saturday last. The foxes have gnawed our tent, and a bag containing biscuit and chocolate. It is usual to put up a piece of rag on the end of a stick to scare them. The Hudson Bay men say, too, that a train of powder laid round the objects is a sure preservative. The sledge covered with gutta-percha runs very well on the snow, but not on the ice, on account of the adhesive nature of that substance. Among our booty are the skins of eight white foxes killed at Port Leopold. Helvetius somewhere speaks of those insects that take the colour of the plants on which they live: is not the case the same with foxes, partridges, hares, and bears? Born among snows, they are white like them. Back again on board at night.

29th October.—Yesterday morning I started again, with four men and a sledge, to bring away the things left at Point Wreck. Leaving two men behind me, about five miles from the point of destination, I joined them again at night, bringing back our buffalo mantles and our tent—precious objects—the loss of which would have been irreparable. We had to pass the night in our tent, in a temperature which, on board, was 10° Fahrenheit. A violent gale sprang up shortly after sunset, and we made ready to start, in case our tent should be blown away. The vapour of our breath rising to the upper part of the tent, and congealing as soon as it came in contact with it, fell back upon us almost in a solid state. This formation of snow went on all night. The same man always had charge of the dogs, and took as much care of them as a good horseman of his horse; he objected, however, to their being admitted into the tent, saying it would spoil them, and that I gave them too much to eat. Though I do not like the manner in which the Esquimaux treat them, I yielded to his remonstrances; but I was excessively uneasy in the middle of the night at not perceiving them outside the tent. Sorely apprehensive of the loss of what was so essential to us, I called out, and what was my surprise to see my favourite bitch, roused from sleep by my voice, come out from under a heap of snow more than two feet high. Slight undulations in the vicinity showed me that the rest of the pack were sleeping in the same manner, without giving themselves the least concern about the snow which was accumulating over them. When I went back into the tent my men informed me, that in the Shetland Islands they let the sheep wander about in the same way in winter, and that they get a very warm covering of fifteen or twenty feet of snow, the heat of their breath making a passage for the renewal of vital air. We had built a wall of snow round three sides of the tent, leaving the opening alone unguarded; but, the wind having shifted, we were wakened by a cold mass that grew larger and larger as it advanced towards us: the snow had found an opening, and a continual reinforcement from without momentarily increased the dimensions of this unwelcome intruder. By daybreak the hills, behind which we had expected to find shelter, were effaced, and all round us stretched a vast plain, levelled by the force of the wind.

The part to be attributed to the latter enemy of ours is, no doubt, large; but I am convinced, from what we afterwards saw, that there fell nearly a foot of snow. We had still some provisions for the evening, but the prospect of another night was anything but agreeable: who could tell how long this might last? We dug out our two sledges with some difficulty; and, as it was as dangerous to remain as to march, we chose the latter alternative.

That ice which we had supposed to be fixed for the whole winter, and which was everywhere eighteen inches thick, had entirely disappeared, being first broken on the coast by the sea-breeze, and then driven to the head of the inlet by the north wind. What thanks we owe to Providence! for certainly, if we had not delayed some days, we could not have thought of reaching Port Leopold before winter. The wind which, outside the bay, blew in the direction of the coast, within it was westward, that it say, full in our faces, sweeping down upon us in such strong gusts, that sometimes men, dogs, and sledges were forced backwards. We could not see ten paces before us, and we should have been lost if we had not been guided by our tracks of the preceding days. In a clear interval we had a glimpse of the ship: it might be a hundred and fifty *mètres* off; it took us three-quarters of an hour to reach her; we had not been seen by those on board. I was the first to jump on board to report our return to Captain Kennedy, who, seeing me panting for breath, covered with snow and perspiration, and unable to speak a word, was frightened, and thought something unfortunate had happened. Our arrival was an agreeable surprise to him, for he had been greatly afraid we should not be able to contend against the storm, having never, he said, seen a more violent snow-drift. My cheeks were frost-bitten, and one of my men had the tip of an ear in the same plight. We had not perceived these accidents, and friction soon removed all trace of them.

How comfortable, warm, and hospitable our poor little schooner seemed to us after these excursions!

30th October.—The days shorten rapidly; and for this long time back, at the hour when the sun attains its highest altitude for the rest of the world, the great ramparts round us cast their huge shadows athwart the ice of the bay. This deprives us of the few days of sun that would still remain to us, astronomically speaking; and as the land stretched south and east of the ship, for the last time to-day its disk has grazed the upper part of the cliffs we see opposite us. I would fain repeat the saying of Diogenes to Alexander—"Get out of my sunshine!" but what is the use of talking to these obstinate stones?

1st November.—We take advantage of the hours of light—I should say twilight, which we enjoy still—to make our last preparations for wintering. We clear the ship of as many things as possible, in order to increase the dwelling-room for the crew and for ourselves, or, at least, to have it as unencumbered as may be. Ventilation is here the most important hygienic element; everywhere—on the bulwarks, the sides of the ship, and in every nook and corner, there forms a crust of ice, which would be tolerable if it remained in the state of ice, but which the variations of temperature at this season often convert into water, keeping up the most unwholesome humidity. A powder-magazine, a forge, and other establishments, are constructed near us, of snow, and give our abode the appearance of a little village of Esquimaux. We are become downright masons—building, building continually, and indulging in all sorts of caprices, the materials being in abundance, and cheap. I have not yet learned how to handle the trowel, that is to say, the snow-knife; and all I can do at present is to cover myself with snow from head to foot; nevertheless, I do not despair of becoming an architect of some ability.

3rd November.—Divine service yesterday, as usual. Tracks of a wolf were seen at some distance from us. Unfortunately, our dogs make those animals give us a wide berth, and I am afraid our hunting-book will not have to record many exploits. Mr. Kennedy talked to me, with an enthusiasm I cannot but envy, of the finger of Providence which, unknown to us, and in the most indirect manner—at least as it appears to us—is leading us to the desired end: to winter in a port on the west coast of the inlet. After having tried the first time to enter here, and left Port Bowen, we tried to enter Port Leopold; and our failure, and our involuntary separation, provoked our lamentations. "And now," said Mr. Kennedy, " here we are in Batty Bay, all in good health and spirits; how can we be otherwise than grateful?" I love and admire this excellent man; so truly pious and Christian, so energetic and devoted. This evening, one of the watch called us on deck, and we thought we were witnesses to a strange effect of refraction, for which we could not account. The edges of the moon's disk were indented, and that in so odd a manner, that we knew not what to think. It was not until we had looked attentively that we discovered an iceberg between us and the moon, the jagged edges of which were projected on her disk, or rather concealed her light from us. How many curious phenomena are just as unreal!

7th November. — Since last Monday we have been exposed to an actual storm, which shakes our masts and rigging with a fury that seems to increase the immobility of our schooner. The snow-drift is thicker, perhaps, than the day we returned from Wreck Bay. As often as the weather allows, our carpenters are employed in constructing a ventilator for the crew's berth, which is almost always filled with smoke and vapour from the kitchen. Two windsails serve to let out these dangerous intruders, and to let in fresh air. A wall of snow is also erected all round the ship and against the sides, up to the level of the deck. Snow being a bad conductor of heat, this measure is intended to preserve a uniform temperature on board. A laundry has been constructed, in order that the men may, in turns, provide for their personal cleanliness during the rest of the winter; for at present it is impossible to dry anything outside, and things hung before our stoves can hardly be made sufficiently dry. However unwholesome this state of things is, our last voyages have forced us continually to postpone the period of this general washing.

8th November.—The weather has brightened a little, and the wind is fallen. I set out with the doctor, in a direction eastward of our mooring, in order to make some observations of refraction; but the last gales have destroyed the ice we supposed to be solidly fixed at the very entrance of the bay, and we are obliged to return without having accomplished our purpose. Our sportsmen have also returned empty-handed, though they saw many tracks of hares and foxes.

10th November.—Yesterday, during a short walk, three of our men found the head of a musk perfectly preserved, the horns still attached; it was, no doubt, swept by the spring floods into the ravine, where it was found. A fox prowls round the ship, attracted by the offal of our bear. To-day Captain Leask and another man killed ten ptarmigans, or rather rock-partridges. At last I have been able to go with Mr. Gideon Smith to the east of the bay, and build a small observatory of snow on a point whence the south is perfectly clear, and where I shall have the sun for some days yet. A black fox has put us all on the alert: the skin is worth £25; but it appears that it is in vain to think of killing them; the rifle is of no avail against the speed of their legs: a trap is the only means of taking them. On our return we fall in with him, and my companion and I pursue him over the rock, to the great detriment of our trowsers, and even of our skins. The cunning brute, guessing, no doubt, that I had nothing but inoffensive instruments with me, made game of us and let us get within a stone's throw of him. In the evening he came back to the neighbourhood of the vessel, visited a trap placed near the powder magazine, and devoured the body of a white fox, leaving us nothing but the head, because it lay within the trap.

14th November.—Notwithstanding the gale, which has again freshened a little, I went again to our observatory, which is three miles and a half from the vessel, and some feet high, to take the altitude of the sun. Since Sunday last the ice has formed again in every place where previously there was running water; it is two or three inches thick, and affords us perfect security at all points. Having been formed in calm weather, it is everywhere quite smooth. This reminds me of the astonishment of Sir Edward Parry, during his Polar expedition (an astonishment I cannot account for in a man of his experience), at not finding ice in the state described by Mr. Scoresby, capable of supporting a wheeled carriage. It is evident that this will always depend on the force of the wind at the time when the ice is formed, and that the state of the days may change in the same winter, and *à fortiori* from year to year. Mr. Leask describes to me the manner in which he has seen white bears taken alive. When the boat has succeeded in cutting off the animal's course in the water, it bears down upon him; a running noose is thrown over his neck, and passed through the block which whale-boats have at the bow; the boat gives way, and the efforts the animal makes to swim tighten the knot. Finally, he is slung alongside, and put into a cage made of a hogshead. The offals of whales, the tail, and the oil mixed with water are the food given him. Mr. Leask saw one sold for seventy five guineas; but the usual price is from £20 to £25. The demands of the Zoological Gardens have raised the price of the article.

12*th November*.—Snow all day. To acquit my conscience I go to the observatory, but return without doing anything; the sun, no doubt, bade us farewell yesterday. The black fox of the other day crosses the bay pretty near; but it is useless to run after him, and our dogs are so stupid that there is no sending them in pursuit of him; at all events, I believe they would find it very hard to catch him. Mr. Kennedy and I set the trap again near the powder magazine. The ship is now entirely covered from end to end, and its sides girt with four or five feet of snow. Two large flights of steps give entrance to the deck—one forward, for throwing snow into the caldron without scattering it on the deck or into the crew's berth, and the other aft. The snow is at present very firm; a little water thrown on the outer parts converts it into ice, and those steps give almost a monumental aspect to our poor dwelling. From the neighbouring heights our little schooner, with its masts rising out of the middle of a narrow block frame, looks like a fly thrown on its back in a pan of milk.

During my observations of the 10th and 11th, I several times burned my eyelids by pressing them against the metal of the telescope. Another circumstance that renders it very difficult to take observations is, that the sun, being near the horizon, its rays dazzle, and the field of the telescope takes in both the luminary and its image. If one colours the image, the horizon becomes obscure, and the observation can in any case be only approximative. Add to these inconveniences, that the observer's breath is turned to ice on the glass and the reflectors.

13*th November.*—Brisk wind and snowdrift again. At last we have caught a prisoner in our trap near the powder magazine; but our black fox turns out to be a blue fox, the fur of which is far less valuable. He is still alive—too much alive for our dogs, who would fain do summary justice upon him; and, as we want to have his skin at least, we kill him ourselves. Yesterday evening Mr. Kennedy put two balls into a box lying near the traps and the powder magazine, mistaking it for the fox. The question is set at rest; the *corpus delicti* exists, and our dogs were unjustly accused the other day of having eaten the white fox. The prisoner we have condemned to death, though he has made no confession, is found to have been the sole culprit. We find evidence of the fact in the tufts of white hair from the fur of his fellow, which he had not time to digest; he has therefore been convicted of a sort of cannibalism, and by the *lex talionis* he ought to have suffered the same punishment; but since we are less cruel than he, and since, moreover, he is dead, we content ourselves with eating him cooked.

14*th November.*—The gale of yesterday is a little abated. Mr. Kennedy goes to the south of the bay, and outside it, to reconnoitre the state of the ice, which he finds broken again, and for the present impracticable. I went to my observatory, but did not see the sun, though I climbed up the glaciers on the north of the bay from fifty to eighty feet higher than the preceding days. In this way we had several times the pleasure of taking exercise *à la montagne russe;* an enjoyment tempered, however, by the consideration that we should have got broken ribs if we had not been able to stop in time. The sky has improved a little in appearance, but nothing has been wanting to these obsequies of the departed sun, neither the sobs of the weeping winds, nor the mournful looks of a day without light. Ossian is truly the poet of the north—of that Nature which moulds the thoughts of men after her own image, and makes them wild and desolate like herself.

15th November.—We have eaten our blue fox, and whether it be that our palates are becoming depraved, or that we were right in our judgment, we unanimously declare him to be equal to the best game we have ever tasted. Mr. Kennedy's trip yesterday had for its object to ascertain if it was now possible to go to Fury Beach; I hope we have not yet given up that intention. This is what I proposed to him: to go next moon to Port Felix or Port Victoria until we meet with natives; to spend the interval between two moons with them; thus collecting any information they may possess regarding Sir John Franklin, studying their manners, &c.; and then to return in January, bringing with us at least two of them; to spend February and the first half of March, the coldest season of the year, on board, during which they would have time to learn a few words of our language; and then to set out again on our spring excursions, employing our guests as interpreters, and after having recompensed them for their services, to send them away not before the end of our tour, in order that they may not make known the depôt of provisions at Fury Beach. But Mr. Kennedy thinks we could not do this, on account of the difficulty of transporting provisions. As for the cold, people sometimes travel in Rupert's Land in a temperature 40° or even 50° below Zero.

17th November.—Though the sun has long disappeared, the twilight still affords us some hours of light: thus, about eight or half-past, we can see to read outside the vessel. It is true that the moon, which we have constantly at this season, helps us a little perhaps: be this as it may, this short respite is usefully employed in completing our preparations for winter. Our various workshops scattered round the vessel would make a stranger suppose there was a little village here. The dear familiar sound of the forge,* the hammer striking the anvil, the shrill noise of the plain, the songs of our washermen, the lively dialogues in broad Scotch carried on in every

---

* An allusion to his childhood. His father had a forge on his premises.

part of the vessel, the movement of the men putting the last hand to our roof—all is full life and animation. It is like a swarm of ants shifting quarters: time presses us, but we are not behindhand.

18th November.—The moon is this morning surrounded by a magnificent halo, which our sailors, habituated to Arctic voyages, consider as a token that we shall have a heavy fall of snow before many days are past. All hands are sent out to hunt. Mr. Kennedy goes to the head of the bay, and sends me with Mr. Smith to examine the first ravine south of the bay. How beautiful is the aspect of those wild cliffs that hang threateningly over our heads! Though the sun is below the horizon, its rays, glancing over the mountains, tinge their snowy heads with sweet rosy hues, which seem incorporated in the snow, and, passing through all gradations, die out in the dark blue of the shadows cast by the indented crests of the great rocks. For the first time, no doubt, the foot of a European disturbs these picturesque solitudes, the echoes of which excite the imagination by their mysterious voices. A rolling stone, the snow creaking under our steps, our very breathing—all resound with a tumult curiously magnified by reverberation, and then dying out in the distance, as if it fled affrighted at itself. The dazzled sight is every moment deceived. Apparently I have plainly before me two men of enormous size; I distinguish all their gestures; I see them load their guns—yes, beyond a doubt they are two of our people coming to meet us; I walk towards them, and fifty yards further I find a stone a foot high, divided into two black halves by a little patch of snow. Night overtakes us before we have marched more than five miles in the ravine and we are obliged to return without ascertaining whether it meets the sea or is prolonged very far still to the south, so as to allow of our passing along it on our overland way to Fury Beach. Our sportsmen have killed in all ten ptarmigans.

25th November.—To-day we have had a little snow; the prediction of the halo is perhaps behindhand with the event, but we wait before pronouncing judgment, in order to be more sure. Two of our men caused us some uneasiness to-day, darkness having overtaken them a little further from the ship than they supposed. Fires lighted outside, and a lantern at the mast-head, brought them back in safety. Mr. Kennedy and I went along the south coast nine or ten miles from the ship. The ice is still in motion, driven southwards by the wind or the current, and we think that we might possibly avail ourselves of this movement as a means of transport, if the ice appeared less liable to be broken during the necessary intervals of rest taken by the parties, which will have to descend the coast.

22nd November.—The same weather. As it is new moon, we have lost the little light that remained to us. Yesterday evening the doctor and I went to examine one of our fox-traps near the ship, but, after plodding all round the place for nearly two hours, we were obliged to return to the ship. To-day we perceive, from our footprints on the snow, that at every turn we passed within less than ten yards of the trap.

The Esquimaux have no God, though Mr. Smith says, that in the environs of Churchill he has heard them speak of another world. They do not bury their dead under ground, because the souls would not then fly off; but they cover them with stones, through the interstices of which the spirits can pass out. The Indians have their arms and all their baggage buried beside them, because it is at night that they go away to the other world, and they could not find them unless they had them under their hands. The law of retaliation seems to be the code of the Esquimaux. In a year when deer are abundant, the Indians kill more than they want in the most improvident manner. This they do from vindictive feelings, on account of the years in which those animals are scarce.

24*th November*. — Our fox-traps have hitherto remained unproductive. In all conscience, we cannot attribute our ill success to our want of skill, for more than half our men are expert trappers; but, as the sides of our traps are of snow, the cunning fox, who knows by experience or by hearsay what a trap is, bores above, or below, or through them, and contrives to baffle us. I took a walk in the first ravine to the north;

it is full of snow, and is several hundred feet deep in some places. It is one of the reservoirs of the inexhaustible cataracts of spring. About the middle, I found a hole, into which I entered, after leaving one of my gloves at the entrance, that it might be known where I was in case of accident. This subterranean conduit, hollowed out by the filtration of the waters through the calcarious ground, communicates with some other ravine; for, though it was quite calm at the surface, I felt puffs of a cold wind that moaned through the cavity, and the cold compelled me to go back. My promenade was arrested at the end of a mile by a perpendicular rock, which I could not even approach; at its foot was a large funnel, formed by the falling-in of the melted snow: its edges even must have been dangerous and slippery, for my faithful bitch, Huske, which had run on before me, began to howl so lamentably when she saw me move in that direction, that I was afraid to go further.

26th November.—Several of our men were sent yesterday in pursuit of game. They saw two hares and three ptarmigans, but killed nothing. The thermometer this morning was below 13°; but what makes us feel the cold most keenly is the high wind that has sprung up. We have remarked since we have been here, that the new and the full moon always bring us bad weather. This morning I talked again to Captain Kennedy about our prospects for the winter, our expedition to Fury Beach, &c.

30*th November*.—It has blown a gale ever since Thursday last, with the usual snow-drift, so thick that we hardly know whether the snow is falling from the sky or is swept up from the ground by the wind. Hitherto the lowest point reached by the thermometer had been 12°, but it suddenly descended to 39° in the course of Sunday. We felt this abrupt transition the more acutely because, as usual, the wind renders the cold much more sensible; and this is the only day on which we have really had some degree of suffering. The fury of the wind increases continually, and the ice breaks with crackling noises, which we sometimes mistake for cries of distress from our labouring masts. All is white around; a white to make one dizzy. After a walk of a few minutes the sight becomes confused, the air seems to thicken, objects lose their forms, and after many a fall one advances only gropingly, and, like the diver, under an element which is not his own.

1st December.—At last the storm has ceased, or nearly so; at least as regards the wind, for the thermometer is no higher than 34°. Mr. Kennedy sends me outside the Bay to see in what state the ice is. The snow first swept, and then heaped up by the wind, is grown firmer, and is now as easy to walk on as the ship's deck. This is one of the conditions favourable for travelling; and, unless furnished with snow-shoes, people must wait after a fall of snow until a great wind has made it firm. The interval of calm during last night has given the ice time to form. The large fragments have been swept away; and, as far as our eyes can reach, we have before us, in all its splendour, that vast white shroud of solid ice. We can now walk safely in almost any direction, and we get over seven or eight miles in a very little time. Once only my foot sank, and then it was into a fissure, or the separation between two floes; but before the water could soak through my canvas mocassins, we spread over them a layer of snow, which absorbed the moisture, hindered the passage of the air as it congealed, and soon formed a complete shoe of ice round my feet.

This is the best, the only remedy against these little accidents which occur at any moment, but which may become very dangerous if one is at any distance from the ship. There is no way of restoring heat to the part that has become torpid; and it is in this way that the frostbites occur, which often make amputation necessary. So we return on board, convinced that the journey to Fury Beach may be easily performed, at least so far as the state of the way is concerned.

For the first time, I made trial of a deer-skin coat given me by Captain Kennedy, and pantaloons of seal-skin. Though wearing little or nothing under these garments, I hardly felt the cold, though it was pretty sharp—32°—the wind blowing with great violence. Woollen stuffs give but very imperfect protection, however thick they may be, the wind passing through them, as our sailors say. It is exceedingly to be regretted that we found no skins for sale at Uppernavick, or that the Esquimaux of Sound's Bay could not sell us any, for our men are certainly not sufficiently provided against cold. When a man is ill defended against that formidable enemy, a very short absence —a distance of only a few hours' journey from the ship—often produces the worst consequences. We are not on the accustomed ground of the men of Rupert's Land, which is very woody, except on the coasts and where the servants of the company travel with impunity at almost all seasons, because they are sure of finding sufficient shelter every evening beside a good fire, round which, after drying their clothes, they lie down, often without any other roof above them than the sky.

If we are not fitted out, clothed, and victualled as completely as we could wish, I now clearly perceive that this is by no means to be charged upon any one as a proof of incapacity or negligence; in fact, the Government vessels are in precisely the same predicament as ourselves. The most erroneous notions are entertained in France as to the information possessed by the English Admiralty. The Hudson Bay Company, which makes a mystery of all its operations, will not allow anything to be published respecting the manners of the tribes in their territory, their resources, or their mode of travelling. Whether it be that the Government has not asked for such information, or has received it in an incomplete form, not one of its naval expeditions has possessed the means of travelling by land; and it is not surprising that each of them has produced so little, since their season of operations has, of necessity, been always restricted to the very short time for which they were supplied with necessaries: besides, even in summer their men, spoiled by a winter passed by the fireside, did not accomplish all that might have been expected of them. In writing down these reflections (as to which the future will show me how far I am in the right), I do not think I am influenced by a feeling of hostility to foreign rivals; but I am struck by the facts. No, I have too high an esteem for Sir Edward Parry and the two Rosses: I do not attack them, but criticise a system, the defects of which nothing could make plain to them, not even their experience in these seas. It does not appear that they ever had with them Hudson Bay men—regular Arctic travellers; they had men experienced in navigation among the ice—good ice-masters—but that is all; whereas in this country, where the ground is solid during two-thirds of the year, a voyage of discovery should be prosecuted by land as well as by sea: and that infers a new series of studies, quite different from those to which a naval officer is accustomed; for then the least details as to foot and head gear, clothes, and food, become of vital importance.

Now, last year Mr. J. Smith was the only man who had belonged to the Hudson Bay service; the greater part of the crew had never been in the ice before, and there was nothing on board of what is requisite for travelling in winter; nay, scarcely the requisites for spending the winter on board ship.

2nd December.—The winter is fairly begun; the temperature leaves us no room to doubt the fact, though we have been gradually and insensibly acclimated. I cannot recollect without laughing those very severe frosts, as I thought them, which I experienced in dear Rochefort, when the thermometer was below Zero, and all the precautions I was obliged to take on my return from the seas of India or Brazil. I find no little matter for reflection in the faculty possessed by this frail machine of ours (not so frail, however, as the poets say in such fine verses) of passing with impunity through these vicissitudes of intense cold and intolerable heat.

I find myself daily more and more attracted towards that religious system which I should call the religion of Nature, because the sentiment is developed in me by the contemplation of the marvels scattered around us by the wonderful Providence which presides over all, provides for all, and foresees all. I lose myself in an inextricable labyrinth whenever I seek to clear up by the too-uncertain lights of my reason those dark, mysterious bye-ways of dogmatic theology, the necessity of which I cannot explain to myself. Whatever be the possible objections to a religion which derives its doctrines only from the innate principles of the heart and the conscience, therein lies for me the basis of all worship—the origin of that gratitude which reminds us of what the creature owes to the universal Creator. My doubts and incessant hesitations, amidst my conscientious inquiries after truth, tell me that I am a rationalist—that is, as some would say, a soul perverted and opposed to the real faith; but who can force himself to believe what he does not understand? And if my poor wearied head, worn with anxiety, happens to go astray in searching for the true way amid the divergent doctrines of Christianity—justification by works or by grace—I always find my mind set at rest when, jumping over intermediaries, I arrive immediately at the conclusion of all religious systems, and address a warm prayer directly to God himself.

3rd December.—In the intervals of calm, copious vapours form a fog, which rises from below the ice on the edges of the bay, in consequence of the difference of temperature; but I have not been able yet to ascertain whether this comes from the temperature of the water under the ice, or from that of the land. Our winter preparations are at last nearly complete. One of our first occupations has been to enlarge the accommodation for the crew, and that at the expense of the cabin, the furniture of which was previously deposited for the most part on the ice. One of the hatchways has been made the entrance to a ladder, the foot of which is outside the bulkhead, so that the outer air may not blow in directly upon the men every time the door is opened. The entrance hatchways, forward and aft, have been covered with a roof six feet above above the deck; the door closes hermetically by means of a cord and a weight; scrapers are placed on deck, that the men may have no excuse for going below with snow sticking to their shoes; and they are enjoined to shake their clothes well before they go below, for every particle of snow is, of course, a cause of damp and vapour. The real enemy, when wintering on board ship in the Arctic regions, is not the cold, against which it is always more or less easy to defend oneself, but damp, which occasions scurvy and rheumatism: every effort should, therefore, be directed to the constant aim of excluding or destroying whatever may cause moisture.

The ward-room is less exposed than the men's quarters to this source of mischief, because the latter includes the kitchens, which, far from being a comfort from the heat they afford, are a continual source of vapour from the food cooked, and from snow which is converted into water. The precautions taken to make the officers' berths wholesome are the same as those before mentioned. The difference of temperature between the air on deck, even under the awning, and in the lower parts of the vessel, is sufficient to condense and convert into ice, at the distance of a few yards from the caldrons, all the vapours that come in contact with the ceiling; so that there is almost everywhere a sheet of ice, especially where there are bolts, iron rails, or plates of metal. The men's berths, ranged all round the forecastle, were not so contrived as to open and allow them sufficient space to move about freely. For that reason we have thought it necessary to enlarge the room; however, we have made but one chamber of it; for the fewer partitions there are the better, since in that case it is so much the easier to ventilate with currents of dry or hot air ; otherwise every angle becomes a receptacle for ice, which is hard to dislodge. As soon as a change of temperature takes place, and the thermometer rises, that ice melts everywhere—not suddenly, but slowly; water trickles down, and must be completely removed, or else it would freeze again as soon as there was another fall of temperature. Wind-sails passing through the deck and the awning serve for the escape of the vapourised moisture, which rises through them by its levity, and is dispersed outside. Another wind-sail is fixed so as to supply a current of air to the kitchen fire, and facilitate the discharge of vapour. The excellent effect of these precautions is in part destroyed by what seems in contradiction with them—the permission given to the men to smoke below. On the other hand, it must be taken into account that tobacco is one of the greatest enjoyments of the sailor, and that, after all, the most serious mischiefs may be prevented by attention to cleanliness. I think, therefore, under very strict conditions, this permission may be continued. Another source of damp, which it is impossible to avoid, is the drying of washed linen and of clothes soaked in snow. *Soaked* is the very word, for the snow lodges on every fibre of woollen garments, so as to make one body, as it were, with the stuff, and wets it as if it had been steeped in water. It strikes me that all the chimney-tubes, &c., might be made to pass through one enclosed space on the deck, which would thus become a drying-room; for everything exposed to the open air immediately becomes hard, whatever be the force of the wind, and tears, or rather breaks, very easily. We have been obliged to put up with this inconvenience, and the officers' things are dried in the after-cabin. It cannot be said that they are much cleaner after their drying, which blackens them with coal smoke;

but at least they are cleared of perspiration and unwholesome elements. Captain Kennedy's cabin and mine, though only some distance from the room in which the stove is fixed, are so cold that at night our perspiration is condensed, and falls back upon us in a fine rain, that passes through our blankets. We have several times had the air in them at 20° centigrade. We should not complain if this moisture remained in the state of ice; but, as we could not without danger sleep all the winter in such watery apartments, we are obliged to shift our quarters; and, as for me, I have a much more comfortable berth. Some degrees of heat—50° Fahrenheit—are quite sufficient, and our consumption of fuel is regulated accordingly. Hitherto it has been found useless to keep the fires burning all night, either forward or aft, and they are all put out at bedtime. The total absence of natural light causes a great consumption of artificial lights, and, as ours consist chiefly of lamps forward and candles aft, it is impossible to imagine how much our eyes and our smell are affected. The smoke of some coal, volatilising solid particles, contributes its share to these causes of discomfort; and we have been obliged to fix tubes, speaking-trumpets, &c., over each lamp, to carry the smoke outside. It is enough to pass near these tubes to be convinced of the quantity of bad things our organs would otherwise absorb. The ice formed in the quarters of the officers and of the crew is taken away as often as necessary; and the nature of this condensation makes me think that it would perhaps be a good thing to have these quarters as lofty as possible: the orlop-deck should be removed when lying up for the winter (supposing the vessel to be a war-brig), and the floor should be lowered; whilst the upper deck would thus become a condenser, which should be cleaned from time to time: the *claires-voies*, &c. &c., might also be converted into condensing-boxes by taking off their tops. I found my cabin partly lined with a woollen stuff, which I was obliged to remove because the vapours stuck to it. The experience acquired, to their cost, by the first navigators in these desolate regions, and the progress of science, have greatly diminished the dangers of scurvy; but one of the very first things which should engage the attention of the leader of an expedition is the diet of his men. The means of resisting cold consist not alone in a judicious economy of fuel or in clothes, but still more in a well-devised system of feeding. The customs of the natives, Esquimaux or Indians, as well as the data furnished by physiology, demonstrate that the basis of that system must be a great consumption of animal substances — of those which contain the most sugar and fat; because, abounding most in carbon, they most accelerate the circulation of the blood—that is to say, the distribution of life.

4th December.—The snow falls in great flakes, and gives a little light, which we still enjoy about noon; during the rest of the day a slaty hue saddens, and makes us feel the cold more sensibly; objects at a little distance from us are all confounded together in one funereal leaden grey. Over head, beneath our feet, all round us snow—nothing but snow. The rugged crests of rock, or the perpendicular faces of the cliffs grinning through it, seem alone to protest against this violation of their nature, and alone remind us that the world is not an immense snowball; and yet there is an indefinable charm in this spectacle, which one feels but cannot express in words—a charm known only to those who have experienced it, because, being before all things men of action, we have not learned to paint what our eyes have seen and admired. When the Indians have no kettle or other vessel capable of withstanding the fire, they heat pebbles, and put them into a vessel of skin among the food or other things they want to warm. A new pipe has been added to our stove, and passes fore and aft through our cabin; on deck, a bed of earth has been laid round the pipe, and a thick bed of snow over the hatches.

8th December.—We have observed, these last few days, that in very calm weather the snow, which is covered with a pretty hard crust, is heard cracking from the effect of cold. This may serve to explain a point which seems as yet undetermined, namely, whether or not the *aurora borealis* is accompanied with a crepitation like that of an electric machine. Our men, who are all from

Hudson's Bay or the Shetlands, affirm the fact in the most positive manner; but how can we set their unenlightened testimony against the assertions of observers who have in vain endeavoured to hear that sound? Yesterday a white fox was run down and killed by our dogs near the ship. A few grey hairs remaining on the tail show that the change of fur is now almost complete. It is plain that the poor animal was attracted towards us by the prospect of a very different reception, for its intestines are perfectly empty. These animals live chiefly on a species of mice which the late cold weather has compelled to hide themselves. Our dogs would not eat their prey. Length from the snout to the tip of the tail, two feet eleven inches; the tail alone, one foot; total weight, five pounds.

9th December.—Traces of the *aurora borealis*—some great white streaks, slightly saffroned, and converging towards the northeast. As often as the thought occurs to me, I deplore the want of instruments that would have rendered possible observations so interesting, for me at least. I perfectly understand why we are without them. Before my arrival, no one seemed to think of making such observations his business; and, besides, Lady Franklin would perhaps have feared lest they should cause the main business of the voyage to be more or less neglected. The Government ships, moreover, were intended amply to fulfil this part of Arctic exploration, and are better fitted for it in every respect than we. In the afternoon, when returning from an excursion, I saw the full moon rise over the northern hills, and I could not help stopping, being struck at first with the absurd idea of a fire—an idea which lasted but a second, but which those great yellowish gleams cast on plains of snow suggest much more naturally than what is called the fires of the sun.

11*th December.*—The storm is raging furiously without; and it sometimes seems as though the wind pierced through our double enclosure of snow and wood. I know not whether this is the cause of the moral discomfort I feel; but, during these last two or three days, I have had fits of savage misanthropy, which make me see everything in gloomy colours. This morning I had for the first time, a pretty sharp religious discussion with Captain Kennedy. We plied each other so hard that we ended in very bad humour. He believes in revelation: for him the Old and the New Testament are of the same authority, and flow directly from the Deity. This I cannot admit. Alas! I have not faith, and my rebellious reason revolts against what it cannot explain. I admit the New Testament as inspired by the Spirit of God; but I do not believe in the prophets, in the Holy Ghost, or at least in the latter otherwise than as a symbolic figure. What is to be done? My conscience tells me I am not wrong: I know well that, in seeking to extract from all systems what is good in them, I should not be supported in my discussions; but, *ma foi!* I cannot help that; I yield to the current of what is passing within me. I perceive that in politics and in religion one has friends only on condition of making certain concessions; but what I have not, never shall have, nor wish to have, is, the spirit of sect: I shall never be able to sacrifice my interests to my sentiments. It is here I recognise the omnipotence of free will—the only real freedom—the only one worthy of man. Why should I attach myself to doctrines which offer me no complete certainty of truth? No, I will never lie to myself, and my mouth shall never say, yes, when my heart says, no. I will avoid among foreigners, as I do in my own country, putting forward even my convictions, when I know well beforehand that few men can be convinced, or will listen to anything but what they like to hear; but when I am questioned, I will reply, at all hazards, in accordance with my conscience. I have hitherto experienced, that whatever relates to metaphysics—to abstract subjects—cannot be demonstrated; one must believe, or not believe. In short, all I could say may be summed up in one short phrase: either a man has faith, or he has not.

The wind springs up again, from time to time, with a force to which we are growing used. The snow, raised by gusts, falls down in such thick showers, that it is sometimes impossible for us to say whether it is fresh-falling snow or a drift. At such times the snow, of course, deepens the darkness; but to-day we cannot sufficiently admire the remarkable luminous effect of the moon, whose disc seems a hole made in the thick vault above us: it is like the light shining into a cellar through a loophole.

13*th December.*—The bad weather at last gives us some respite, and this is the finest day we have enjoyed for a long time—if indeed that interval of time may fitly be called day in which the only light we have is that of the moon, and the sky all sparkling with stars, and which not long ago we should have called a fine night. Alas! what is become of truth, if it is no longer to be found in proverbs, which are said to be the wisdom of the nations? Poor nations, whose wisdom refuses to believe in stars at mid-day! About one o'clock, at two several times, we hear a dull rolling noise, which can only be compared to the sound of thunder, or of a wall tumbling down. It is certainly not thunder, for that is rarely heard in these climates, as all our whalers tell me; besides, we are past the season of tempests. It is in spring, too, that the ice-rocks fall asunder; and the only probable supposition I can admit is a movement of our chains under the ice. Messrs. Kennedy and Anderson, who have witnessed earthquakes, are of opinion that the sound is the same as that which they then heard; it agrees also with the description given me by several officers who were in the West Indies at the time of the earthquake of 1848.

This is Saint Adelaide's day, and my thoughts are in France, at Rochefort, with that good mother whose *fête* it is. Since I began my roving life, eleven years ago, I have alway been abroad on this anniversary. Remembrances of my childhood come to me during my reveries, carrying me among that group of children, delighted to kiss their beloved mother, who weeps, I am sure, for my absence, with my sister, my dear Adelaide! Poor mother! What uneasiness I caused her before I entered the navy, by the fears my turbulence excited; and afterwards what new anxieties for my life! Why can we not live the past over again? How obedient, respectful, and industrious, I would be! Poor dear, excellent mother, to whom

I owe all I know and all I see, may I yet, by my assiduous attention, sweeten the latter days of your life, which hitherto have been almost constantly passed in tears and uncertainty for the morrow! Do we ever know what trouble and tears we have cost our mothers? May God hear my ardent prayers, and may those dear friends guess my thoughts, and feel in their hearts the kisses I send them from far away!

16*th December.*—A new shipmate is born to us: my bitch Husky has got a pretty little pup, which the crew have already christened *Arctic*. The subterraneous noise of yesterday evidently had reference to this unexpected birth, and that explanation may doubtless contribute to relieve the minds of our superstitious Scotch, who maintain that such noises, for which no cause can be assigned, bode nothing good. Since the completion of our labours outside, all hands have been employed in making snow-shoes, of which every man is to have a pair. Some prepare the wood; others cut the thongs, or are beginning the nets under Captain Kennedy's directions. These snow-shoes are not without elegance:

we are making two kinds of them; one round, like a racket; the other handsomely curved in front, like a Moorish slipper. In the afternoon there is generally a school-class, in which the more erudite, acting as monitors, expound the complications of arithmetic to their less-instructed shipmates. It is perhaps not uninteresting to note, as evidence of the development of primary instruction in Great Britain, that of our whole crew, consisting of men who all began betimes to work for their daily bread, there is only one who does not know how to write. I believe that in Scotland primary instruction is more general than in England. From time to time we come on deck to admire the lustre of these splendid nights in the Arctic regions, where the sky is so abundantly set with stars— "those everlasting flowers of heaven," as Basil the Great says—and is coloured by the fleeting brightness of the *aurora borealis*.

19*th December*.—Mr. Hepburn, talking to me to-day of Lady Franklin, tells me that in Van Diemen's Land she bought extensive tracts of land, on which she established settlers, defraying all the first expenses for them, and furnishing them with the means of cultivation on such conditions that in three years some of them were out of debt and in a thriving way. I am every day more filled with admiration for Lady Franklin's noble character and superior mind. After despatching the *Prince Albert* in 1850, she spent the remainder of the season in the Shetlands, recruiting colonists for Van Diemen's Land, where most of those poor creatures, who were almost starving at home, may in a short while become prosperous landowners.

22nd December.—At last we are arrived at the shortest day, and now we shall approach the sun more and more; not that we have had total night, yet even at mid-day the darkness has always been such as to enable us to see a great number of stars; but this very day, about half-past eleven, a reddish band to the south, stretching eastward, served to indicate twilight. The winter hitherto has been very mild, and when there is no wind, a temperature of 20° or about 30° centigrade below zero is by no means disagreeable. Old Æolus is our only implacable enemy, who stops only from time to time to fill his bags again; for, since Sunday last, the wind has blown fresher than ever, and we have been several times in fear for our awning.

25th December.—It was new moon on the 22nd, and to-day a high tide lifts the ice above its usual level. The noise heard on the 15th is repeated to-day, added to which we have that of the ship, which is every now and then jerked and shaken. This movement of the ice explains to us what we did not understand the other day. My own notion is that, the ship being lighter than usual, the chains are more strained in the water; and that it is the movements of the *Prince Albert* that occasion these earthquake like sounds.

Christmas, which is not observed by the Church of Scotland, is nevertheless a day of rejoicing, like our New Year's Day. A departure is made from our habitual dietary in favour of the crew, who have a little firewater served out to them, and soon exhibit a degree of merriment which it is pleasant to be able to produce so easily. This resource, prudently employed, is not to be despised in regions where resources are so few. The present breach of rule is a judicious indulgence on Captain Kennedy's part; but it is impossible not to attribute to his temperance system the harmony, good humour, and mutual desire to oblige, displayed by our ship's company. Nevertheless, they are by no means possessed with a strong repugnance for ardent spirits, or a blind faith in the merits of cold water. But are not their incessant toils and hardships, and the detestable ideas in which they grow up, a sufficient excuse for a weakness that counterbalances so many good qualities? Besides, who knows but that an indulgent physiology may, at a more advanced stage of the science, find a reason for the propensity in an exclusively salt diet, and even in the exhalations of the sea? Let him who is without sin cast the first stone at them; those who command them have other duties. In this case, as in all others, prevention is better than cure. Sir John Ross attributes to a similar system the absence of scurvy during the long winter of his five years' detention (1829 to 1834), and our own experience seems to prove, that in these climates the dietary should exclude spirituous liquors. Jack Tar does not approve of a complete reform of his habits, especially in what concerns his darling sin; no more do his brethren across the Channel, for the sailor seems to make nothing of the differences of character created by geography; he is a sailor before he is a Frenchman or an Englishman; but, once more I say, he must be made better in spite of himself. As regards his personal interest, the sailor is a minor who must submit, and whose prejudices must not be listened to.

28th December.—The sky has been generally clear these last days, and this evening we have the first time a complete *aurora borealis*, or *northern lights*, as our Shetlanders call them (they also call them *dancing lights*). Great luminous rays like the Milky Way, but with a slight yellowish tint, divide the vault of the sky, issuing from the zenith, from which they spread like the leaves of a palm, widening at the base. I do not know that mention has anywhere been made of this singular phenomenon. We are about to quit our purely defensive attitude in a few days, Captain Kennedy intending to repair with three men and myself to Fury Beach, to examine the state of the provisions. Though the want of deer-skins is a great obstacle, we have not to regret the loss of time in preparing our equipments for the fine season; for, during the three months we have passed here, we have been exposed, as it were, to one continual storm. The ship's company can now enjoy a little recreation, and the popular game of football puts them in high spirits. We have all been struck by the facility with which one is put out of breath here, although the barometer does not indicate a great pressure; and we should, perhaps, have attributed this effect to want of exercise, did we not recollect that the Americans had occasion to make the same remark during the whole of their winter. After all, however, the sailor must be made to bestir himself; exercise is here the grand secret of health, and nothing must be neglected to overcome the aversion to movement which possesses men usually so active, the chief cause for it being a change of temperature of 60° or 70°, and often more.

This indolence goes still further. The first colds produce a moral torpor—a somnolence of mind, which in my own case showed itself principally when I wished to write. In reading over my journal, I often find English words, which are there only because I should have had to think a little in order to find the corresponding French words, and because of late the English words have become more familiar to me.

CHAPTER V.

1852.

1st January.—THE wind and snow-drifts, which hinder us from enjoying some games in the open air—our only recreation—thrust us back upon the reflections that occur very naturally at this season, and we involuntarily compare the day with what it generally is in our homes. Without seriously regretting our present situation, we cannot help turning a melancholy look on the past. We all entered with ardour, and of our own full accord, upon the sacred cause in which we are now engaged, and not one, I am sure, thinks of counting fatigues or privations, or looking backwards; no, it is forward, to the future, that our eyes turn; but the recollection of our native land, and of all that is dear to us, instead of weakening our spirits, gives them a fresh impulse. For my part, I passed the whole day, and the nights preceding and following it, in rummaging all the nooks and corners of my memory in search of some new detail I had forgotten of home friendships and affections. Dear good friends! if there exist between sympathetic beings those influences which magnetisers talk of, you must know how your names are all joined in a fervent prayer every night, and how by turns you possess my whole thoughts before I lie down, and in my waking moments at night. Our poor sailors do not, perhaps, enjoy in an equal degree with us this facility of retracing an agreeable past; many a countenance among them looks rather lengthened by the absence of that sovereign elixir which on Christmas-day so gladdened their honest faces.

Where was I this day last year? Where shall I be next year? The past, alas! we often know but too well; but the future always seems to us more smiling and full of promises, which it does not always keep. For my part, I have reason to thank Providence for having endowed me with a high degree of confidence in the future. In that way only will I make use of predestination; and, as I told Mr. Kennedy this morning, when my conscience points out to me the end, I adopt that idea very readily; nevertheless, I perceive how dangerous it may become in certain cases. Courage, then, and welcome to the new year! I shall see what will become of my hope.

3rd January.—The two days have been spent in preparations for our going to Fury Beach. The weather is fine, and the moon will be full on Monday; we shall perhaps have a prosperous trip, though I think it will be longer than Captain Kennedy supposes. As I should be greatly vexed that the least thing be done and I not there, and just because of the repugnance our men manifest for travelling at this season, I shall be delighted to show them once more that a French officer will never hang back, but, on the contrary, is always eager to be foremost. The east wind has brought us a rise of temperature of 10° to 15°, and we find it really very warm, with the thermometer at 4° below zero. Mr. John Smith, who was long in Hudson's Bay, is the person who makes most objections, on account of our scanty equipment. We will take the tent with us, though it will encumber us a little; but I suppose Captain Kennedy has resolved to do so, and I quite approve of it, that we may not be quite at the mercy of the caprices of Mr. John, who is our only builder of snow-houses —I mean, the only expert one. A fox has been killed by our dogs not far from the vessel: the poor brute is so lean that it is evident it was in quest of food; it ventured into a dangerous neighbourhood. This is one of the contradictions that astonish me there, where so few are found. How is it that these animals are left by Nature without resources against hunger? All the ships that have wintered in these regions speak of the great number of foxes attracted by the emanations from the vessel, and it is probable that our dogs alone keep them aloof. Mr. Leask says he has found holes in which foxes had deposited game in reserve. The weather is clear and quite favourable, but we shall not set off before Monday, in order to observe Sunday.

4th January.—A second fox has been found on the ice, dead of hunger, and reduced to a mere skeleton. Two others prowl round the ship in the course of the day, attracted by the scent of the meat which has been exposed to clear it of salt; they are so tamed by hunger, that we could almost kick them before they go off reluctantly, frequently looking back at the forbidden fruit. Fortunately for our hungry visitors, our dogs are asleep, and they are safe from us on Sunday.

11th January.—Here we are once more at home, all safe and sound, after an excursion to Fury Beach.

CHAPTER VI.

EXCURSION TO FURY BEACH.

12th January.—Messrs. Kennedy, John Smith, W. Miller, W. Adamson, and I, set off on Monday, the 5th of January, with an Indian sledge and four dogs, taking provisions for several days, our camping materials, a tent, and a box of pemmican of about ninety or a hundred pounds. A part of the crew accompanied us out of the bay; and, though the ice appeared broken at a little distance, the mild temperature (20° Fahrenheit) promised us an easy journey. No sooner, however, were we left to ourselves than we encountered greater difficulties than we had expected, in the unevenness of the ice and the resistance of our dogs, who found themselves for the first time harnessed to an Indian sledge. The Esquimaux harness them two and two, and the traces are merely thongs of leather passed round the neck and the body, whilst the Indians fasten them one by one. Our sledge, too, is one of those which are called *flat*, composed of long planks, curved up in front so as to form a pretty wide arc: a cord passing from one end to the other serves to give it spring. With its sixteen inches of width, and its twelve feet of length, it is so flexible that, under a load of about five hundred weight, it glides over the snow from one block to another without danger of being broken, where the Esquimaux sledge would soon be smashed to pieces. It has also the advantage of not sinking in the snow where it is still soft; whereas the Esquimaux sledges would sink to their full depth in it. We reached a distance of ten miles before dark. Our tent was soon set up, and a wall of snow three or four feet high raised round it for shelter. Sleeping in a tent with the thermometer at 24° may seem rather uncomfortable, but the absence of wind was in our favour, and the fatigues of the day had so well prepared us, that we all slept too soundly to think of the bad weather that might come upon us. The sky was overcast when we awoke; the wind had sprung up from the south—that is to say, from the quarter toward which we were to march; and as the twilight failed us, on which we had reckoned to resume our route, we had to wait for the rising of the moon. At that season, the moon remains a very short while below the horizon; and we should have had light much longer if we had travelled along the coast, which is everywhere flanked with cliffs more than two hundred feet high. As we expected, the vapour of our breath had covered the walls of the tent, and came down upon us in a little fall of snow whenever the frame was shaken. It was as a precaution against a sudden change of temperature that we had taken the tent, for after a long march it might be impossible for us to construct a snow-house with sufficient speed, that job occupying an hour and a half of hard work, rendered the more difficult by the snow, which flies about and gets into the eyes, mouth, throat, up the sleeves, and everywhere.

Four or five miles further on we found on the hard snow impressions of human feet, so distinctly marked that they caused us some anxiety; for if the Esquimaux had come so far north they would certainly have passed by Fury Beach, and consequently that depôt of provisions, so necessary to our operations, would exist no longer. Our fears were soon set at rest, however, by a fresh discovery made further on—remains of fire and three cases of preserves, which would infallibly have been picked up by Esquimaux who had come north. It was probably one of Lieutenant Robinson's camping-grounds. This relief to our apprehensions gave us fresh vigour, and at ten in the evening we halted, having travelled a dozen miles in the day. The sky was overcast, and foretokened a fall of snow before long; and as, after all, it was prudent to ensure ourselves a refuge for the return journey, we built a spacious snow-house, our dogs being left, as usual, to their own industry to provide themselves with suitable quarters near us. A piece of mackintosh spread on the snow, a blanket for each person, and two buffalo-skins for the five of us, formed our customary bed; and our repast was a bit of pemmican, with a little tea, which we thought delicious. This was a luxury in which we ventured to indulge, because our trip would not be a long one, and we had been able to carry a little fuel with us to boil our water; and seldom has a good appetite given a better zest to any meals than to ours. The difficulty of travelling in this region is so great, that the weight to be carried must always be carefully considered; and the arms of the party, a few things for change, and the provisions, always prove, in the end, a more cumbrous mass than is desirable.

We had marched seven or eight miles on the third day, when Captain Kennedy, who had gone ahead to reconnoitre, came back and told us that decidedly we were beaten, for the ice was so broken at the foot of the cliff that there was not the least possibility of our passing it with the sledge. We had surmounted so many obstacles on the two preceding days that we did not imagine there were any which could stop us; sometimes hoisting up our sledge with difficulty to the top of an ice-block several yards high; sometimes rolling, men and dogs, down a hillock formed by the collision of two floes; most frequently groping along a route hardly discernible by the dubious light of the moon; informed of the direction we ought not to take only by the frequent falls of our guide, but at the same time not knowing the safest direction; sometimes squeezed between the sledge and the rocky asperities of the iceblocks when our dogs turned too short, or falling into a hole, and dragged over the snow before we had time to get on our feet again.

We were compelled, however, to yield to evident necessity, and an excavation in the snow offering us a house half made, we completed it in a pretty durable manner, in case we should be obliged to wait for more favourable circumstances. After some hours' rest, it was decided that Messrs. Kennedy, J. Smith, and I should endeavour to reach Somerset House without baggage, and that the two other men should wait for us in the snowhouse with the dogs; our object in this excursion having been, above all, to examine into the state of things at Fury Beach, and see if fresh visitors had been there since the visit paid by Sir James Ross's detachment. The very imperfect indications we possessed respecting this part of the coast, and our uncertain estimate of the ground we got over every day, left us in almost complete ignorance of our distance. We took, at all risks, a day's provisions with us. At nightfall we arrived at the foot of a high precipice, which we guessed to be that described as being three or four miles north of the spot where the *Fury* was wrecked in 1825. The coast sinks gradually from that point, and several ravines which we passed in the dark, whilst they helped us to identify the ground, were many times near being fatal to us. In fact, it cannot be conceived how uncertain and fallacious are the appearances presented by the snow even at night; all objects are confounded in one uniform hue; difference of surface disappears; outlines are lost; one gropes about in a semi-transparent fog; the eye, wearied by continual attention, loses the power of distinguishing anything; the foot rises to surmount a rise in the ground, and comes down upon vacancy; the ground seems to stretch out horizontally before you, and suddenly you roll down a steep hill. Our sticks alone gave us warning when this danger was imminent; and after a long circuit, we found that we had come to the edge of a dried torrent. As in autumn, the ice was always most disordered at all projecting points; and there the hummocks rise in some places to a height of fifteen or twenty feet; some, which were the growth of several years, were even double that height. Sometimes we had to make our road over the points of frozen snow or glaciers by cutting steps with our hatchets, where we could pass on foot, though not quite without danger, for there was water below, and the consequence of an immersion would infallibly have been the death of one of us. Several times we thought we perceived the object of our quest in the distance: a flat stone of small dimensions, or a projection of rock had been mistaken by us for the vast tent that so long sheltered the shipwrecked crew of the *Fury*. Some fragments of wooden and iron cases, however, showed us that we were not far from Somerset House; and, recollecting the sketch inserted in the narrative of Sir James Ross, we soon arrive there.

We raised a shout of joy, but no one replied to it. The roof of the house, formed of a mast-top covered with the running gear of the *Fury*, was still entire. Lieutenant Robinson's report speaks of a cairn surmounted by a cross, which we did not find. Only the bears and the foxes had broken through the canvas wall of the poor dwelling, which for us, however, was as sumptuous in that snow desert as the greenest oasis in an ocean of sand.

One of our hopes was dissipated, for though we never supposed—at least, I did not—that Sir John Franklin had left any memorandum here, we had thought, in spite of ourselves, that perhaps one of the vessels of the Arctic squadron might have sent some document to the spot. But no; everything was in the same state as Lieutenant Robinson had described; we could not even find the papers he left there in 1849. It is true, our scrutiny could not be very extensive, as we had no light with us. To the sadness caused by this disappointment was superadded a feeling very easy to imagine, especially by sailors who have seen the remains of a shipwreck.

When the *Fury* was forced by the ice upon

the beach to which she bequeathed her name, everything that could be carried away was taken out of her, but the greater part of the rigging, sails, and anchors, were of necessity left behind. The provisions were placed as far as possible out of the way of harm from the sea; and though Sir John Ross's abode of a year made a pretty large breach in them, the beach is still literally strewed with fragments: here high piles of wooden and iron boxes, of all forms and sizes; there barrels of flour and salt meat; further on, the ship's anchors and grapplings; then two damaged boats, oars, and boat-masts—all of them things of inestimable value, as the least experience tells a sailor, especially in regions where nothing of the sort can be procured, and of which a large portion will no doubt be lost, since we cannot carry them off with us. It is some satisfaction, nevertheless, to have put our hands on these resources, which have become doubly valuable to us since we have been unable to touch at Navy Board Inlet; but we cannot help pitying the poor Esquimaux of Boothia Felix, for whom such a great store of iron and wood would have been a mine more precious than the richest gold and silver mines for us.

A pretty keen appetite, and a natural desire to know in what state of preservation these provisions were, made us open some of the cases taken at random. They were all frozen, but quite sound; and the vegetables we tasted had lost nothing of their savour in the interval of thirty years since they were embarked. We recollected Sir John Ross's remarks as to the facility with which Herculaneum and Pompeii would have made known to the dietary of the Romans, if they had possessed the art of preserving their viands. A small barrel of lime-juice was but very slightly frozen, and melted immediately at some distance from a fire we kindled. After having refreshed ourselves, and carried our researches as far as possible, we went back about midnight towards our snow-house, delighted with having at least some good tidings to bring to our companions. Captain Kennedy had decided that we should rejoin the ship, and return with the rest of the crew, which could be accomplished now that we were sure of subsistence at Fury

Beach. At five we reached our encampment, the guardians of which, uneasy at our absence, had determined to set out in search of us at daybreak.

The wind had turned against us, which made our journey back a little more disagreeable than that outward. The stones rolled down from the rocks three or four hundred feet high, and the blocks of snow, sometimes as big as hogsheads, showed us that it was not very safe to keep too close to the cliffs, so we generally remained on the ice; but almost everywhere the unfrozen water flowed at a distance from the coast, the blocks of ice drifting southwards. It is a very remarkable circumstance, which Parry was struck with in 1825, that seven days out of ten the ice is swept southward. Whither does it go? Is there an issue at the bottom of the inlet to Brentford Bay, or even to Pelly Bay. (By the bye, many persons doubt that Dr. Rae, who never obtained his longitudes otherwise than approximately, has been very exact in general, or very justifiable in not examining the bottom of Pelly Bay.) What added to our discomfort was the state of our clothes, all drenched in snow, which the heat of our bodies melted at night, and which, freezing next day, made them heavy, hard, and cold, as if they were made of lead.

Our Hudson Bay men are less accustomed to excursions in these naked regions than in woody countries, where every evening a good fire, fed with fuel abundantly supplied by the neighbourhood, enables them to warm themselves and dry their accoutrements. The certainty of this resource leads to negligence, and they do not always take sufficient care to shake off the snow which insinuates itself through all the openings of one's clothes. They are also very ill clad for our situation, their clothes being of wool, which is notoriously well adapted to retain every particle of snow, especially when that snow, fine and dry as the dust of our highways, flies in clouds before the least breath of wind. The safety of the traveller and his very life are secured by a multitude of precautions, which do not seem puerile to those who can appreciate their utility. One ought not, for the sake of excluding the cold, to cover himself with warm stuffs, which immediately bring on a copious perspiration, which turns to ice at the least lowering of temperature, when he stops to rest or ceases from violent exercise. In this respect, deer-skin garments have a great advantage, for they are light and at the same time impermeable to wind; but that same impermeability is a cause of perspiration; and so, when we felt our hands moist, we took off our gloves, until a fresh sensation of cold forced us to put them on again. We did the same with our headgear. In spite of these precautions, we had every evening to dry our shoes and stockings, and our gloves, which we could only do by laying them on our breasts or under our armpits. This is not wholesome; but it is the only method, and it is easy to conceive how rapidly rheumatisms occur under such a practice. The nose and the mouth are the parts of the face most sensitive to cold; but it is hardly possible, nor would it indeed be very prudent, to cover them, on account of the vapours that escape from them. Those among us who muffled their faces in cravats, nose-covers, &c., could not take them off at the day's end. Do what we would, our blankets were always frozen; and on Saturday morning, though we were two days' march from Batty Bay, we resolved to make a bold push, and not halt until we were on board the *Prince Albert*. Though we were not greatly fatigued, that was the day on which we had to suffer most. The north wind lashed our faces, and it is impossible to imagine the sensation it causes when the temperature is at 30° below zero (centigrade). The pain seemed to us like that of being scourged with thongs of leather; each blast, in fact, seemed to bear away pieces of the epidermis. This smarting is succeeded by a numbness of the skin, during which the affected parts become bluish; if unhappily they whiten, it is all over with them; they are irrevocably frozen. We were obliged to stop every now and then to examine each other's faces, for under intense cold, especially if one has reached the stage of numbness, one does not feel that he is frostbitten. For my part, I paid for my noviciate by more numerous frostbites than the rest, which among the Indians or the Huskis would have exposed me to the derision of the girls or of the wags of the country. Experienced persons know that when they feel an itching, they are frostbitten; and in that case one must not hesitate to take off his gloves, and rub the affected part well with his finger. It is a disgrace for a man to let himself be frozen, and the young Indian prides himself on coming off safe and sound from exposure to cold. A handful of snow applied to the part is a sure means of restoring the circulation of the blood.

When the gust was too strong, all we could do was to turn our backs to it; but the snow got into our throats and nostrils, and between our eyelashes, often gluing our eyelids together, so that we had to pull out the ice. We thought to find a protection from the cold in our long beards, but the snow always stuck in them, and, condensing in large masses of ice, fastened them to our clothes; and we thought ourselves very fortunate in having scissors to get rid of the encumbrance. The Esquimaux are generally beardless; and, if we had reflected more on the ways of Nature, we ought to have arrived *à priori* at this conclusion. The same inconvenience attends every kind of cover and muffling for the face, and the best measure of prudence is to habituate the skin to these low temperatures.

We set out at two in the morning, intending to make a short halt on the way; but we were barely able to melt a little water and hold a morsel of biscuit in our torpid fingers. We could not help laughing many a time at the strange grimaces each of us made on putting a tin can of melted snow to his lips. He always withdrew it with roars of pain, and sometimes, it must be owned, with imprecations. The first touch of the cold metal causes no extraordinary sensation, but presently one is reminded of the law of the equilibrium of caloric, by a sharp pain caused by the tearing off of the skin, which sticks to the edge of the vessel. At five in the evening we were at the entrance of the bay, rolling and tumbling head-foremost among icebergs, grounded on the reef, and steering our course as well as we could for the ship in darkness increased by the fog, created by the vapours rising from the clefts between the floes. Passing the foot of a berg where the snow was saturated with sea-water, we observed the same phosphorescence as in October at Point Wreck; but this time the phenomenon was evidently due to the water itself, and not to the remains of fish. Our arrival surprised the crew of the *Prince Albert*, who did not expect us so soon; and once more we enjoyed the exquisite comforts of a good fire and good beds, dry and warm. What horrible sufferings wretches endure who are cold and hungry! Why are not certain of our legislators sent for a few winter months to travel in the Arctic regions?

During this excursion we saw tracks of only one bear (but they were old) and of two foxes, which followed our march from our third encampment to Fury Beach, besides two great black crows, which our men would not have killed for any consideration. We left a hundred pounds of pemmican at our third encampment—the tent and its posts at the second.

CHAPTER VII.

ON BOARD.

19th [?] *January.*—WE have to-day the strongest gale of wind we have yet encountered. What thanks do we not owe to Providence for our preservation! for in the state in which our blankets and our equipments were, it is hard to guess what would have become of us if we were abroad. We cannot too much congratulate ourselves on our arrival, or own how often we have erred in our judgments, and been led by the hand to port. Nothing gave us reason to expect the approach of bad weather, and certainly we should not have harassed ourselves with fatigue to arrive at once, had it not been for the wish to spend Sunday on board. The accomplishment of this religious duty has perhaps been the means of saving our lives.

The whole ship's company is actively employed in preparing for an excursion on a larger scale than was at first intended; and in making up clothes, they avail themselves of our experience of last week, which showed us how little protection European garments afford against cold and snow. We have conformed to what should be the rule in all countries, and adopted as much as possible the native style of dress, which, being closed in front, gives no admission to snow, and has a hood, which may at pleasure be drawn quite close round the face. Our men, who are, perhaps, a little too well aware of all we want, feel some repugnance to such an excursion at this season, ill provided in all respects as we are. A pair of mocassins lasts but a few days, and we are to set out with two pairs each. Captain Kennedy, who did not know the country, and reckoned on finding here deer-skins, &c., had assured them that they should be supplied on board with everything requisite; and now they reproach him for this, saying that otherwise they could easily have provided themselves.

It is under such circumstances, when it is necessary that every one should be able to bear with the situation in which he is placed, and contribute to the general welfare and success by his own intelligent and cheerful efforts, that the advantage of a superior education becomes clearly apparent. Oh! my God, how I thank thee for having enabled me to obtain that education where so few poor children can receive it! The more I advance in my career, the more I am impressed with this truth; and I wish I could form a crew of officers exclusively. I should be sure of achieving all that is humanly possible.

16*th* [?] *January*.—Still the same gale. We are beginning to recover from our frostbites, which the warmth of the vessel has rendered more apparent. Our faces are covered with spots like bruises; the skin peels off in flakes as large as halfpence, and the new skin is so tender that the part is almost immediately frostbitten again. With this exception, we feel no other effect of our excursion than a violent appetite, which we can hardly satisfy. This is one of the bad consequences of these prolonged bodily exercises: everything in us becomes materialised, in spite of ourselves.

Physical fatigue kills thought; and I must own that I have several times caught myself dreaming, with my eyes open, of a cup of coffee, or a piece of new bread and a slice of ham.

16*th* [?] *January*.—But our desires cannot range very far, even in our palace of the *Prince Albert*, without our finding out the impossibility of satisfying them. Last year's provisions, unshipped at Aberdeen, and shipped again this year, were taken no care of in the inverval, and are now quite damaged. The salt butter, pork, &c., are all nearly unfit for use. One of our casks of lime-juice got frozen, and lost all acidity. We shall soon run short of everything, though having depôts at Port Leopold, Fury Beach, and Navy Board Inlet, because it will perhaps be impossible to reach them.

17*th January*.—The hurricane has ceased at last, but for how long? The atmosphere has a lustre of which we have no conception in our southern climates. Those hills, covered with snow that looks as polished and hard as the whitest marble, seem to start out from the horizon; their outlines are marked with incredible precision, and their smallest details stand out in black relief against the bluish sky. The twilight to-day gives us more than six hours' light, and it is with great glee we accept this prognostic of less sombre days.

Nothing is more irksome than this life by the smoky lights on shipboard: the artificial glare is fatiguing, and causes a weakness of sight of which we are all conscious. I very much fear the effect of spring on my eyes; but otherwise all are well. Poor Mr. Kennedy is the lamest of us all; I fancy he has always had a contempt for physical pain, and yet, at five and thirty, he is crippled with rheumatism.

What bad nights we spend very often, when, worn out with fatigue, we do not take time to build a snow-house, and lie down covered with snow, wet, without even caring to change our damp clothes, and we spend the night in shivering! we must all turn on our sides at once, and rub our backs and feet to keep us at all warm. The next morning no one is rested, and every one is cross.

According to the usual custom on shipboard, some more delicate provisions had been provided for the use of the cabin; but we have shared everything with the crew, and always followed the same diet, for which I heartily applaud Mr. Kennedy, for, if we share the privations of the crew, they will be less likely to hang back from other difficulties.

Our experience, although short, has proved to me that whatever a man can endure, I can endure. It is my settled conviction, that will and moral energy can always take the place of physical strength, and I trust to emerge honourably from these trials: thank God! I was not reared in luxury.

19*th* [?] *January.*—This is the stiffest gale we have yet encountered; these dull winter days follow and resemble each other in a terribly monotonous way. Although we are pretty well inured to cold, we have been unable to go out, the wind is so sharp. The little schooner, comparatively so warm and habitable, quivers beneath the violent squalls; she is almost torn out of her bed of ice; and it seems as if the wind was determined on rooting up our two masts, which, as they stand with their yards amid the snow, look like the leafless branches of trees in winter.

21st January.—The gale has at last slackened, and such is the difference of sensations caused by the wind, that, although the thermometer has fallen from $-10°$ of Fahrenheit to $-26°$, we feel as if the temperature had risen agreeably. We are all busy with our travelling preparations. Next Monday is the day appointed, and, notwithstanding our state of destitution, there are sufficiently good reasons for hastening our departure. The ice is broken by every gust of wind; but, with this temperature, a few hours' calm is sufficient to freeze it thickly enough to enable one to get across the various openings. The winter having been tolerably mild, the melting of the snow will be early, and increase the difficulties of our return; moreover, and for the same reason, the icebergs in the bay will be early in motion, on account of the torrents of fresh water poured in by the numerous ravines surrounding it. We shall take the entire crew with us, and the ship will want all her hands; there will be a large quantity of ballast only transportable over the ice, as we have but one boat. Our wanderings over the hills will be less difficult, as the snow is now hard, and presents greater facilities for locomotion.

We must not lose sight of the principal object of the expedition. We shall not be favoured in exploring still unknown regions, as we shall have but little sun; but our researches can be made with the greatest facility, and this absence of prolonged light is what will preserve us from the danger of snow-blindness during part of the time that we march towards the south. We have certainly not come here to divert ourselves and be comfortable; and I, for my part, applaud Mr. Kennedy highly for his decision and intrepidity.

We used to think that about this time of year the inlet would be frozen from shore to shore; but it will not be so this year at least, and we once more bless Providence; for, if we had wintered at Port Bowen, our expedition could not have been made this month, and the year would have been entirely lost to us, as all research would become impossible, or at least extremely perilous.

22nd January.—The thermometer fell this morning to − 44° Fahrenheit, one degree higher only than the Americans experienced last year. Either our mercury is not very pure, or our thermometer is inexactly graduated; for a certain quantity of this metal put into an earthenware vessel, but placed on the waistcloths, and receiving perhaps a certain amount of warmth from the wood of the ship, only froze at − 42°; and the pain caused by its touch was intense. At − 39°, it was almost in a melting state, and by shaking the vessel it became perfectly liquid. Our northern men consider a very brilliant *aurora borealis* last night, and another one to-day at five o'clock, as prognostics of dry, cold weather. The wind is never very high, it is said, when the temperature is very low; but this evening, at − 35°, we had a very rough gale from the north-west. The method by which we shall pursue our survey southward is still undetermined : Mr. Kennedy talked the other day of dividing our party into two detachments—one on the eastern, the other on the western, coast of Boothia Felix.

25th Jauuary.—Still the same gale blowing. A hungry fox shamelessly enters our larder; I set the dogs on the trail, and, carried away by my eagerness, thoughtlessly follow them. The sharp cold warns me of my imprudence; and, although I have not been more than five minutes out, I return on board with my nose and cheeks frozen: the skin peels off as if I had been scalded.

26th January.—We must postpone our departure until next Monday, because we are not ready, and the weather is not sufficiently settled. The wind having fallen, Mr. Kennedy and I go out to examine the state of the ice. It is still broken at a little distance from the shore, and travels to the south, presenting a splendid, perfectly smooth floe, of which the ice, impregnated with seawater, is wet and slushy, wets the feet through in a very few minutes, and renders the hauling of boats extremely difficult. The prospect is not encouraging, and it is but little likely that we shall accomplish all that we wish; but I agree with Mr. Kennedy that we must attempt it. It does not seem likely that the inlet will be frozen over this year, which increases the chances against us; and it was probably the same thing last year. For my part, I think that, after deciding to set forth, we should close our eyes, and not look back until we are far on our way.

Mr. Smith has made a pair of wire spectacles for each of us. A case has been made for our second chronometer, which does not, fortunately, seem to have been disturbed by our three months' repose. It will not be possible to have exact Greenwich time, on account of the great variations which all our instruments have undergone in the voyage: the one that Mr. Kennedy wears changes its time daily, according to the longer or shorter period it has been out; but they will be very useful for taking observations.

A fourth change of plan. This time it is decided that we shall all start together, the seamen and bad walkers accompanying us only as far as Brentford Bay. We have reckoned that, after passing Cape Bird, the Magnetic Pole, Cape Felix, Franklin Point, Montreal Island, Pelly Bay, Port Felix, and then ascending the northern coast, we shall have made a journey of 1400 miles in the space of four months. We shall take good care to conceal this from the men, for fear of alarming them; for never was such a journey undertaken at this season of the year, especially by people as ill supplied as we shall be. We must trust to Providence for food and clothes!

I delight to find in Mr. Kennedy that nobleness of nature I so love and revere, and that ardent enthusiasm which alone can overcome difficulties. I long for some change which shall make me forget the vexations inseparable from life in common, in this dreadful region, amongst men of such dissimilar education and ideas.

27th January.—The preparations for our final journey, and the thousand little things of which one does not feel the necessity until the last minute, leave us no rest. This is one of the great advantages we have over the other expeditions, one of whose greatest difficulties was the providing employment for the men during the winter time.

The captain is anxious about the ship, and

that is easy to understand: it is the natural result of that division of employments which exists amongst us. I am trying to impress the necessity for taking one of our *cloak-boats* with us; otherwise, if we should have to cross the two or three chains of lakes which divide the two seas south of Port Victoria, we may meet with the greatest difficulties.

In spring, or rather as soon as the sun has appeared, the melting of the snow forms inundations which may close up the paths; the lakes overflow, and, meeting over the slightly elevated ground which divides them, form a small ocean, which might stop us for a long while. In addition, as the season will be earlier than usual this year, we may be stopped at the Magnetic Pole, or on the peninsula of Cape Felix, by the unexpected breaking up of the ice. We should therefore, in every case, be guilty of extreme improvidence, and have no right to make any complaints hereafter, if we were not to avail ourselves of every resource in our power.

On the whole, and considering the state of the season, I think I should have adopted the following plan:—To send depôts of provisions from Fury Beach to Brentford Bay and Cape Bird; to explore Brentford and Cresswell Bays, as well as the west coast from the places visited by Sir James Ross in 1849, to Cape Bird; to send for the boat we had left at Fury Beach, and the *youyou* left at Cape Seppings; and to bring these boats down to Brentford Bay, if there is a passage open; or, if it is possible, to hoist them overland on the other side.

The rocky coasts are always the first to be free from ice; therefore we may hope that it will be possible to steer a boat along the western coast of Boothia Felix about the month of April, and we could then do in two days what it would take us ten to do on foot.

I do not mean to say that the execution of this plan would present no difficulties; we all know too well the fate of plans quietly arranged by the side of a good fire, with plenty of food within call. But I foresee the obstacles which the melting of the snow, fatigue, insufficient food and clothing, and snow-blindness, will oppose to our journey during the fine weather.

In any case, we should be quite sure of

not losing our boats, whilst at present two of them are in danger. However this may be, and as I hate nothing more than being perpetually doubtful of success without ever attempting anything, I am anxious to be off.

Foxes are always roaming about near us, and we have unfortunately set up no traps, because our dogs destroy them; otherwise we should have made use of the ingenious method employed by Sir James Ross at Port Leopold. As these animals traverse great distances, every fox taken alive was set free again, bearing a copper collar announcing the presence of the ships and the provisions deposited in that place.

30th January.—During a clear moment, and notwithstanding threats of snow, I have carried four cases of pemmican (three hundred and sixty pounds), and about fifty pounds of coal, in advance upon our future road.

The weather was as mild as it always is when the sky is laden with snow; but, on the other hand, the snow is so wet and yielding, that we had to perform the journey twice; so that a distance of eight miles became one of twenty-four. For the first time for a long while, the ice adheres to the shore, and is pretty smooth as far as the eye can reach; but it also is so slushy that three men and our five dogs could hardly drag two hundred pounds' weight along on the sledge. Such are the minor troubles which make all journeys here so laborious; and it may not be uninstructive to see on what a slender thread our existence hangs.

If it is windy and cold, we are frost-bitten; if it is not so cold, we cannot walk. How is it possible to do anything with the elements conspiring against you?

The track of a hare was found on the spot where we deposited our burden. Some tufts of yellowish grass, preserved probably by the snow, prove that this spot must be a meeting-place for herbivorous animals. On my return, I am told that partridge tracks have been seen on the land.

31st January.—I went to my November observatory, hoping to discover the sun, which ought to make its appearance to-day, if the refraction is the same.

Although the atmosphere was pretty clear,

the horizon was bathed in a dark mist, which concealed the land to the east, and which the rays of the sun are as yet unable to pierce. But above this mist a few clouds, richly tinted with purple, and long trails of light converging towards the horizon, indicated that the sun was not far off. For the last few days, a light has glimmered in through our lenticular glasses, and rejoiced the interior of our cabin. This thrifty, not to say stingy, dispensation of natural light, nevertheless revives us, and we feel restored to life again.

2nd February.—A great quantity of snow has been falling ever since yesterday. The wind has become easterly as well as higher, and this causes a corresponding rise of the thermometer—snow being a bad conductor of heat. This is a fresh vexation, as we cannot now set out until another gale of wind has either blown this snow away or drifted and hardened it. Our impatience is all the greater, because it is indispensable for us to know in good time whether our plan of operations is possible, or whether we must change it entirely.

I am now convinced that we shall find some of Sir John Franklin's men amongst the Esquimaux of Boothia Felix, or at any rate some trace of their passage there. If the men lost their officers, they would probably prefer remaining with the natives, having no idea of the distances or the formation of the coast, and being ignorant of the position of Fury Beach, and of the fact that boats had been left at Port Elizabeth.

5th February.—Snow has fallen for the last two days, and this is perhaps the first real fall of snow that we have seen this winter. What strange beings we are, and how well Bossuet might have added a chapter, headed " *Des Voyageurs Arctiques,*" to his " *Histoire des Variations !* " During the last month we have done nothing but lament the interminable gales which kept us here; and now our only cry is, " A gale! a gale! —my kingdom for a gale!"

I am very much out of humour on account of an accident which is irreparable here. The pocket chronometer I carried fell down, stopped, and we are thus deprived of half our resources; but, as we shall only take a sextant with us on account of the weight, a second chronometer would have been of no use but as means of verification.

7th February.—Several of our sportsmen have gone in quest of ptarmigan tracks; but the snow, which is in some places several feet deep, deprives of security all but those who have snow-shoes. One of the latter killed four partridges out of the six he flushed. The return of these birds, or rather their reappearance, looks like a favourable omen; I say reappearance, because it seems that the partridges, at least, pass the winter buried under the snow, and in a probably torpid condition. Several burrows were seen by Mr. Leask in Wolsternholme Sound. One or two large crows have also been seen. What puzzles us is to know what they feed on. The ptarmigan's stomachs contained the buds of the dwarf-willow, or of the dwarf-birch.

Mr. Kennedy saw the sun yesterday from the top of the hills east of the bay. We have verified that the thickness of the ice free from snow is five feet in the bay close to the ship.

10th February.—One day it snows, the next

day the wind blows, the following day it snows again; so that if this continues we are likely to remain here a long while. Three foxes have been caught by our dogs in the last few days, and their success gives them a spirit of which they seem proud; the result of their chase is mostly announced by themselves: they leap about us in a state of excitement if they have performed some feat during the night, and lead us immediately to their victim.

The poor foxes are clothed in their white winter fur; their black eyes and noses alone seem to project from this perfectly spotless coat, and there is nothing more graceful than the movements of these animals (when they are not exhausted by hunger), as they sport with the efforts of our most active dogs, and distance them in a few bounds. A strong north-west wind sweeps down the snows amassed upon the hill-tops, and the torrents which roll over the cliffs to the south of the bay form a cataract of enormous thickness, which falls at a great distance from the perpendicular wall.

12*th February*.—The hardened snow splits,

and emits a cracking sound of an evening, quite independently of the *aurora borealis;* indeed, I believe the *aurora borealis* has nothing to do with this accompaniment, the origin of which is not known. The various explanations given to me seem all the more strange, because, unlike every other physical question, the cause has been sought before the existence of the phenomenon was ascertained. Reading, dancing, our artist Mr. James Smith's violin, and the organ given by Prince Albert, constitute our usual evening amusements.

14*th February.*—We have returned from a little excursion which nearly cost us dear. Yesterday morning Mr. Kennedy, Mr. Anderson, the carpenter, Andrew Irvine, and myself, set out with the intention of carrying forward another portion of our provisions. The weather did not look very fine, but precisely on that account, and not to seem to draw back, I joined the party, which Mr. Kennedy himself commanded. The snow, still very yielding outside the bay, made the road very fatiguing as far as the little depôt I had formed the preceding week; but the ice being strong and whole, we added our provisions, and carried two miles further five cases of pemmican (four hundred and fifty pounds), six gallons of spirits of wine (seventy-five pounds), and four muskets; as much, in short, as our dogs could draw. The gradual darkness and the increasing thickness of the snowdrift had been warning us for some time that our return would be no easy matter.

We had no sooner set our faces to the wind than we were all violently frostbitten; fortunately the wind fell a little, and, rubbing our faces incessantly with snow, we slowly began our road towards Batty Bay. By the time we reached the south point of the bay it was quite dark. Having eaten nothing since the morning but a piece of biscuit without water, for fear of losing time, we were all very exhausted, and lay down by turns on the sledge. Fearing lest we should not be able to proceed any further, we proposed covering ourselves over with a buffalo robe (only one between five of us), or lying down in the snow as best we could, or returning to the spot where our provisions were; it was, however, decided that the best plan was to continue our road.

When we left the south side of the bay, the darkness was so intense that the opposite shore, only a mile distant, became invisible; and the wind, changing every minute, ceased to help to guide us; so that we wandered at random until, in a momentary gleam of light, the Polar star showed us the direction we were to take. This helped us to reach the north coast of the bay; but, once there, we could not distinguish if we were to the east or west of the ship: even on the shore, at the foot of the lofty hills which surround it, these very hills, buried in snow or hidden by the drift, were undistinguishable; and after following for a short time, which appeared a very long one to us, the direction which we fancied the best, we had to retrace our steps, on discovering that we had got out of the bay.

We had still to face the gale: our dogs, worn out with fatigue, lay down, and although we let them loose, they would not move or attempt to guide us; perhaps they had lost their way as completely as we had. The men who made up our party were half beside themselves and dejected at the uncertainty of our position; everything combined to make our prospect anything but a pleasant one. Every five minutes we stopped to rub our faces and melt the snow which stuck our eyelids together; the stones of the beach, which we did not dare to leave for fear of losing our way, cut our feet to pieces. Poor Mr. Anderson slipped every moment from one stone to another, and we were obliged to lead him by the hand. Fortunately, those of our dogs who were loose found one of their previous tracks, and, followed by those still harnessed to the sledge, set off at a gallop, proving that we were at last in the right path, which revived our courage. Following the windings of the bay, we reached our powder magazine, which we should have missed, though only a few yards from us, had it not been that an oar, the colour of which stood out against the snow, called our attention to it.

The ship, at two hundred yards from the powder magazine, was not visible; but, by forming a parallel line along the shore, and keeping in sight of one another, we got on board at last, and were warmly welcomed and congratulated by all our men, who were very uneasy on our account. It was ten o'clock, and, as it was about five when we reached the other side of the bay, we had been five hours wandering and turning about the ship. They had been the more alarmed on board because the two dogs arrived about nine o'clock, bringing with them no note or indication whatsoever of our position.

Ourselves once in safety, we began to think of our poor dogs, who had been, no doubt, kept back by the sledge, which probably stuck fast in the icebergs on the coast; but the storm was too violent for a man to be sent out of the ship, our strongest light being invisible at a distance of twenty yards.

All night the wind blew with increasing vehemence, which told plainly enough what would have become of us if Providence had not guided our steps. Fortunately, this morning, during a temporary lull, the dogs were found stuck fast, as we supposed, and so entangled in the traces that they could not get on. We could not help envying the facility with which these poor animals bear the rigours of such a climate.

Thank God! both men and beasts are now safe and sound; the former, however, obliged to acknowledge their inferiority by the numerous marks they bear on their faces and hands. Notwithstanding this slight inconvenience, we cannot be too thankful to that Divine mercy which protects us through these various dangers. To be in the least distant from the ship may at any moment become fatal; and these little excursions are even more dangerous than a real journey, in which you are provided against every emergency.

Andrew Irvine nearly fainted before reaching the south point. The doctor tells us that three of our companions were so despairing, that they were repeating their prayers as they went along. This morning, having started with four men in search of clothing which had fallen off the sledge, I recognised the spot where we landed last night, quite close to the heap of stones piled up on the north point, about three hundred yards from the ship.

After such undertakings, the desire most keenly felt is the desire for sleep.

We can hardly help laughing when we look at our grotesque, swollen faces and the bruises, which make us look as if we had been fighting. The doctor feared, for a moment, that Mr. Kennedy's nose was completely frozen. I am, thank God! the least disabled one of the party, owing to the constant attention I paid to keeping every exposed part constantly warm, not fearing even to take off my mittens in order to do so; and I am proclaimed an experienced traveller! A little intelligence and moral force soon give this experience, and make me hopeful for future expeditions. After all, we are in the hands of Him who watches over all his creatures; but we must help ourselves a little.

Our men have spent the winter a little more confined than perhaps was right, but not without being employed in work rendered necessary by our state of destitution. The consequence of this prolonged imprisonment has been to make them more susceptible to cold: it is possible to familiarise the epidermis to this low temperature, even if it cannot be made quite invulnerable; but, do what you will, especially with the sailors, they always manage things in their own way, and their clothing, exercise, and health can never be too closely watched, even at the risk of their finding this watching troublesome.

We have found on the snow the recent tracks of a bear going southwards, and have seen two crows. The foxes had opened our sack of coal, and tried to break open the pemmican cases; but it is probable they burnt themselves, for sheets of tin stick to the skin like glue.

15*th February.*—Mr. Kennedy is kept in bed by a violent inflammation of the cheek and eyelids. When the snow which gets under the eyelids melts, and gums the eyelashes together, they get pulled out; and this infallibly produces inflammation. At ten minutes to eleven, the sun has for the first time made its appearance southward of the bay, hailed with hearty cheers; its whole disc has risen above the hills. We had not seen it since the 30th of October, on board the vessel; so that it remained hid one hundred and eight days from the *Prince Albert*, though we ought to have seen it yesterday, and even the day before, if the weather had been favourable. Its rays are still without heat, more so, I could almost imagine, than the moonlight of Mozambique, if natural philosophers did not assert the contrary; but they are not the less welcome; and I now perfectly comprehend the worship paid to the great luminary by certain tribes, and the festivals in its honour instituted by the ancient Scandinavians.

18*th February.*—I went with six men last Monday to build a snow-house at the south point, and left there means for kindling a good fire. Such a shelter would many a time have got us out of difficulty, if it had been sooner constructed; and I think the measure generally commendable: it would be a useful exercise for the men at the beginning of winter. Remarkable effect of vertical light.

19*th February.*—The plan of our intended excursion is changed. Captain Kennedy thinks it will be impossible for us to accomplish the expedition on foot; and it is at present decided that we shall only go to the Magnetic Pole. I think this an advisable change; only it is proposed to return by the same route, then to take the boat or boats, go back to Brentford Bay, and launch them on the lakes of Boothia: but I propose, in going for the boats to Port Leopold, to compass North Somerset; then, descending by Cape Walker, to explore the lands seen by Sir James Ross in 1849, and to return to Brentford Bay after having pushed on as far south as Victoria Land and the coast eastward of the Coppermine River. According to my plan, we should go on foot as far as Cape Isabella, and return by Port Felix, Victoria, and Elizabeth: in that way we should soon know, at least, if any of the unfortunate wrecked men were among the Esquimaux.

20*th February.*—We collect near the ship about thirty tons of stones for ballast. The vessel being lightened by our winter consumption, the pressure of the ice will tend to heave it up more and more; and then it is evident that, being empty, it would be less able to resist the outward pressure which may take place when the thaw comes. Our Esquimau sledge, the under part of which has been smeared with a mixture of oatmeal porridge and snow, does wonders on this smooth ice; and the dogs haul nearly a ton weight, without difficulty, at every trip.

22nd February.—Yesterday we had a little gale, during which the foxes came about us as usual in search of shelter, and because, no doubt, they had scented further off the savoury exhalations from the ship. One of them was killed by us, and another by one of our dogs, which ran it down in a few minutes. The weather has improved, and the gleams of light that lately were confined to the summits of the hills, now flood their slopes exposed to the south. The sun chases the shadows that yesterday disputed the ground with him, and yielded only step by step: everything wears a smiling aspect; and, but that we miss the songs of birds, we might fancy ourselves in more favoured regions. All is glittering round us, and no one thinks of remarking the absence of all vegetation.

Thermometer at noon, on board in the shade, $-5°$ Fahrenheit; in the open air, $-26°$.

23*rd February.*—We are making our preparations for to-morrow. Only six men will accompany us; the indolent and the convalescent will not join us till some days later; and as, after all, this determination was necessary, I rejoice that it has been resolutely taken. I am impatient to see ourselves entered at last upon the more active part of our expedition, and an end thereby put to all the petty vexations of a life so full of fears and torments of all sorts.

Zodiacal or vertical light, with a remarkable parhelion in the sun's vertical, surmounted by two arcs like horns, and joined to the sun's disc by a perpendicular train.

28*th February.*—After being detained last Tuesday by bad weather, we at last set off with our two casks and our five dogs, intending to stop as long as necessary, to carry away all the provisions we have deposited at various times on the coast. We encamped the first night about thirteen miles from our ship. There I collected, on Thursday morning, the several deposits between that point and Batty Bay; and at ten o'clock, under Captain Kennedy's direction, we carried the whole nine miles further, and returned to sleep in the same snow-house. On Friday morning I go on board to take charge of the rest of the men, and lead them to Fury Beach, where Captain Kennedy expects to have arrived some days before me. Though the north wind blows fresh in my teeth, I set out the more gaily because Captain Kennedy has asked me if I was willing to return *alone;* and as, even among experienced travellers, it is considered very imprudent to go any distance whatever, except in groups of at least two, this confidence in my address makes me rather proud; but on arriving on board I make no boast of having let one of my thumbs be frostbitten. Immediately on my arrival the doctor tells me, what I knew already, that Captain Kennedy had left sealed instructions, in case of his death, transferring the command to me, and a letter addressed to myself, containing a medal. It is one of those which Lady Franklin had struck for the expedition; and already, in November, Captain Kennedy assured me he could not begin the distribution better than with me, &c. &c. I am much touched by this attention on the part of this brave and good commander.

Some of the men left on board have not recovered the effects of our expedition on the 13th; others are suffering under obstinate dysentery; and Adamson, our dog-driver, has a toe frozen. Captain Kennedy himself has his face still covered with scars; but assuredly he is not a man to hang back, and I know not who could complain when he thus sets the example.

We saw a crow on the day of our departure: these birds are our only faithful companions during the whole winter. On what do they live?

29*th February.*—Our preparations are completed; and if the gale, which bellows more than ever, will but give me a moment's respite, I shall still have time to take the last provident measures to-morrow morning. A few hardships will soon have broken in our novices; and the best way is to push them forward, sure as I am that, once in the midst of the stream, they will strike out vigorously.

I have succeeded in banishing from my mind all anxiety for the future. A glance at those dear written souvenirs of France, a hearty prayer in the evening, and then come what may. My days are numbered, and nothing will happen without God's permission.

In the course of the day the thermometer rose to $+22°$ in the sun, and $+10°$ in the shade; so that the snow covering our tent fell in incessant drops of rain: the interior of our snow-houses is also covered with water. What a fortunate year it would be for a ship in search of the north-west passage; and how glad I should be to be on board one of the vessels of Collinson's expedition, with Captain M'Clure! But here the temperature is a bad chance the more for our pedestrian journey.

2nd March.—I have been detained these two days by a gale of wind from the north, which leaves us not without uneasiness as to the discomforts of poor Captain Kennedy and his men. I hope, however, that they have made for Somerset House; otherwise I fear the effect of such a beginning on the *moral* of our men. Their stock of coal was exhausted when I quitted them, the foxes having eaten the sack I had carried forward in advance; but, luckily, they have plenty of spirits of wine. A cold diet very soon causes cramps in the stomach; besides, it is impossible to get water otherwise than by melting snow. This is a fact of which people unacquainted with these barbarous regions have no notion, and they imagine that the snow is as easy to swallow as in our comparatively mild winters. The enormous difference of temperature between the intestines and the snow, or the outer air, causes a sort of suffocation, or rather a sensation of intolerable burning.

The sun is already very powerful, and the reverberation from the snow distressed us all on the 26th, though we took great pleasure in observing the splendid parabola of rainbow hues, marching before us like the fiery pillar before the tribes of Israel. I am quite put out to see so little enthusiasm in some of our companions, who seem not to have foreseen that a service like ours could not be accomplished except by surmounting a great number of obstacles, and enduring some hardships. O! if they all had Captain Kennedy's energy and resolution, we might do a great deal.

If Captain Kennedy would take my advice we should use all our people only as carriers, to form a large depôt at Brentford Bay, by small portions taken successively and from point to point along the coast. Then taking two of the best men and our dogs, we should proceed as well as possible to the accomplishment of our task. I am decidedly of opinion that small detachments only are capable of acting, especially with dogs; for a dog does not need so much food as a man, and wants neither fuel, nor cooking utensils, nor bedding, nor shelter of any kind; so calculating that each man would carry or drag a weight equivalent to that of his own baggage, and that a dog draws a hundred pounds on a sledge, which is the average; our dogs would draw five hundred pounds of food, which, at the rate of nine or ten pounds a-day for five dogs and four men, would make it possible for us to travel thirty days, reckoning from the last depôt; and as the weight would diminish faster than our strength, it is certain that the thing could be done; whereas with a party of twelve men the force is expended in drawing camping and other materials, and four or five hours are requisite for preparing meals and suitable quarters.

In fine, God grant us his aid! We have need of it.

30th May.—Here we are, once more on board, the exploring party in good health, relatively speaking, and all safe and sound, or nearly so. Unfortunately the case is not quite the same with those we left behind us; they having suffered from scurvy. Thank God, however, there have been no deaths; and now that we are all together, and our spirits raised again, it will not be long before the invalids are completely cured. For my part, my heart is full and overflowing with gratitude to Him who has preserved and sustained us in our various perils, and who has saved me, doubtless, in order to return me to my family and to the happiness of embracing those who are so dear to me. (Psalm ciii.)

31st May.—We ourselves have not been

exempt from scurvy, but have been attacked by it in a slight degree compared with those we left on board. The causes have evidently been the bad state of the ship's provisions, the moisture that prevailed during the winter, and, above all, the want of sufficient exercise from October to March; but, fortunately, the malady is not without a remedy, and we may hope that Providence, who has hitherto protected us in so signal a manner, will not forsake us. Of all our invalids the one who gives me most uneasiness, and whose share of sickness I have often during our absence wished I could take, is excellent old Mr. Hepburn, for whom I feel the greatest respect and veneration; but I am sure that the anxiety which was long felt by all on our account has had the greatest share in his illness, and that now he will quickly recover. Knowing the small quantity of provisions we had taken with us, most regarded us as men they were never more to see, especially as a party of four men who returned four days ago from Fury Beach, did not announce our coming. But at last here we are, arrived from Port Leopold, a point directly opposite to that from which we were expected. A truce, then, to our respective inquietudes; and let us think of nothing but rendering thanks to the Creator of all things, whose guardian hand has raised us as often as our tottering feet have stumbled.

During our absence I regularly kept a journal every day, which I here insert.

CHAPTER VIII.

EXPLORATION OF THE LAND.

4th March.—THE weather is not much better, but every day that separates us from our companions seems to me lost; and, as I do not think we can count much on finer days at this season, I have determined to make a start. We shall find quarters ready made along our route; and, that part of the labour being saved us, I think we must take the rest as it comes. Five sailors (Magnus M'Currus, Linklater, Irvine, Adamson, and Kenneth, the carpenter), R. Grate, the boatswain, and Mr. Anderson, the third officer, form the little caravan placed under my orders. Two sledges contain our baggage and a small stock of provisions. Superior portion of a halo, and a horizontal parhelion. Reached the first camping-ground after marching eight hours.

5th March.—Fine weather; wind from the south. After ten hours' march, reached the second camp station, situated thirteen miles from the first. I find a little note from Captain Kennedy, telling me he has been detained there by the last gale, and that we shall find on the road a depôt of five cases of pemmican, four gallons of spirits of wine, and muskets. A parhelion without a halo.

6th March.—Fine weather. The refraction is such at the rising of the sun, that the vertical diameter of its disc appears to be but half as long as the horizontal diameter, thus giving it the appearance of a very elongated elipse. My men, who are somewhat fatigued, cry out beforehand against taking with us the additional burthen announced to us; but I have found, as under many other circumstances, that it was better to say nothing to them of what I intended to do: and after reaching our third encampment, situated six miles from the second, I had some tea made; and when they were somewhat cheered by the refreshing beverage, I explained to them that this job would have to be done sooner or later, and that to postpone it would only increase its difficulty; the day, too, was favourable; and, in fine, I had the satisfaction of seeing all our provisions within one day's march only of Fury Beach. This encampment is that at which we were stopped in January; and, after breaking in through the roof, we found the house we had cut out of a bank of snow in so dilapidated a state that we had to make fresh excavations. Our work was somewhat like that of miners; fortunately, our ground was not hard, and came away easily enough without blasting. A parhelion and portion of a halo.

7th March.—Fine weather. At one in the afternoon I had the pleasure of shaking hands with Captain Kennedy, whom I found, as well as the rest, in good health. We were not looked for before the end of next week, and he complimented me on my activity and zeal. They were detained several days at the second encampment, where, after exhausting their spirits of wine, they were obliged to burn the tent-posts left there in January. An avalanche, which would have crushed them under its fragments had it been nearer, fell at a little distance from their encampment. With the exception of frostbites, which must be expected at this season, they are all in good condition. At the place where we halted in January, and at several points on the coast, we had to cut steps with our hatchets in the banks of snow, which we could not have passed otherwise, their faces were so steep. Somerset House appears to be no comfortable asylum, at least in its present state, and the three days they had passed there appeared to them very cold. Fortunately, Captain Kennedy does not give them time to grow torpid, and will certainly show them the way to keep warm with work. No one has visited these places since ourselves; and this time the notice left by Lieutenant Robinson has been found, on a post at some distance from the house, where the darkness had hindered us from seeing it.

8th March.—We are all employed in rebuilding Somerset House, which is not so easy as in 1832. The word *house* would lead one to expect a shelter of some sort or other; but the fact is, that the thing so called —once perhaps appropriately—now consists of nothing but a framework of spars, covered with a rotten and tattered sail, which would have been blown away piecemeal but for a heap of ropes lying upon it. Fortunately, the house was made large enough for a residence of several months; and, although we are almost as numerous as was Sir John Ross's crew, by contenting ourselves with half the space, we find scraps of sails enough to enclose the portion formerly reserved for the officers. A wall of snow from six to ten feet thick, rising to the roof, barely serves to secure us from a blast that seems to make a sport of tormenting us; and we have to heap coal on coal in the two stoves which have been found and repaired, before we can feel anything like warmth. Fuel is, fortunately, abundant; ten tons of stone coal, remains of barrels, masts, &c., lying on the beach.

A double halo, the outer arc being distinct to an extent of only 30° on each side of the horizontal diameter.

Sum total:—The state of the provisions found here is such that they may furnish a reserve against the return of those who are to go forward. A considerable quantity of preserved vegetables, soups, flour, and sugar, some barrels of lime-juice, and several barrels of dried peas, are perfectly sound. Unfortunately the ropes, many of which have been left lying on the ground, are no longer fit for anything but to make oakum. Lances, harpoons, and other fishing utensils, small shot, grape, knives, articles of barter, &c. &c., testify the solicitude with which the *Fairy* and *Hecla* had been fitted out, and the care with which, amidst the bustle and anxieties of a shipwreck, Sir Edward Parry had everything landed which he could not carry away. As I observed in January, Sir John Ross owed to this care a considerable stock of provisions in 1829, and the existence of his crew in 1832—33. Of the two mahogany boats which lay here, one is quite decayed and unserviceable; the other may be repaired, after being well caulked, for the oakum has started out of all the seams. This seems to be an effect of dry cold, which acts on the wood in the same way as excessive heat in temperate regions.

We have made ourselves beds, in which we lie two and two, by forming a sort of frames, to which we fasten a network of ropes; but the smoke is such that we are forced to cut a hole in the roof, which is far from contributing to our comfort.

Several of our men are ill, or say so; one alleging an incurable affection of the lungs, another rheumatism, or pains that nail him to his berth. Captain Kennedy thinks he detects in these latter too manifest a want of good will, and a wish to shirk the evils of our excursion. He warns them that he will take with him none but men who are resolved to do their duty, according to the measure of their strength; but that nothing can change his determination; and that if he and I are left alone, alone we will march to the accomplishment of our enterprise. I was not present at these explanations, but I am very glad to find that Captain Kennedy knows how determined I am to aid and support him with all my efforts.

Not one day has the weather hitherto been favourable for observations, and yesterday, Wednesday, it blew strongly from the south, the thermometer at $-20°$ Fahrenheit. I was setting to work when, on taking up my sextant, I saw in the mirror that my two cheeks were frost bitten all over, and thereupon I pitched the observations to the devil. It is certainly easy, seated in a comfortable arm-chair by the side of a good fire, to arrange plans of expeditions, and then when they come back, to criticise what has been done, and complain of the paucity of the results obtained. But send these fine gentlemen here, set them to work, and let them say whether the smallest things, which at a distance seem so easy to accomplish, are not attended with difficulties enough to rebut the most patient and well-disposed.

12th March.—The provisions found here consist solely of vegetables, and many of our men are beginning to feel the effects of a diet exclusively of that kind. On the other hand, we cannot touch upon our reserves already, without running short for our excursion. Captain Kennedy has, consequently, resolved to send me on board with five men (G. Smith, M'Currus, Webb, Grate, and Linklater), two sledges, and four dogs, to get a further supply of salt provisions, pemmican and biscuit; the biscuit found at **Fury** beach being too damaged to be fit for use. After a rapid march of five hours we reached the second encampment.

13*th March.*—Fine weather; cold. We halted at the first encampment to repair it, and at six in the afternoon we arrived on board, having this day marched five and twenty miles.

The bear-skins and other specimens left by Ross or Parry's expedition are all in the most deplorable state.

During our absence, the four men left on board have had enough to do in drying the damp berths of our companions and the cabin. What will be the effects of the constant humidity in which many of us have thus lived? What I have read in accounts of voyages, and my own experience, make me apprehensive of them; but the mischief, whatever it be, is done, and it will be time enough to think of it when it appears. They have been visited by a bear, which must have taken notice for some time of the meat placed on a platform outside the vessel to clear it of salt. Gun-shots, fired at sundry times, were powerless against the temptation of a savoury meal;—pigs' feet and quarters of beef, master bear!—and when the animal could do no better, he came, several days in succession, and lay down on the ice, not far from the ship, to feast his eyes upon the tempting meat. Our men watched for him, and on the night of the 8th of March they saw him approach stealthily, and spring with the agility of a cat upon the scaffold, which was more than four feet above the snow. Their guns were ready this time, and a terrible howl told that the robber had been punished. It was impossible, and perhaps dangerous, to pursue him in the dark; but next day a long track of blood, and the marks of a broken leg and of many falls on the snow, showed that he had received a mortal wound. A number of foxes were upon his track, no doubt promising themselves an abundant quarry. Such is the effrontery of these animals that one day, whilst the bear was lying on the ice gnawing the remains of his thefts, one of them came up in play, threw itself upon the bear, and did not run till it had received a stroke of the paw and a significant growl.

One blue and two white foxes have been killed. The latter are beginning to drop their winter coat, and some grey hairs appear through their white fur. Our men have taken advantage of the absence of the dogs to set fox-traps, but hitherto without success.

14*th March.*—Light breeze from the southwest in the morning; clear weather; in the afternoon calm. At eight P.M. the thermometer is at 46°, and one of them asserts that at ten o'clock he saw it mark 48°. It is calm, and the cold felt outside the ship is by no means piercing or disagreeable. One whole day is spent in cleaning, washing, and other preparations for departure; for, though hitherto I have made it a point scrupulously to observe Sunday, I think it right to take it on my own responsibility to break the laws of the Sabbath under our present circumstances. I must here remark, that the English constitution recognises and prescribes the right of repose on Sunday for everybody; nevertheless, I have encountered no opposition on the part of the men placed under my orders.

16*th March.*—The bad weather has again

disappointed me, and delayed our departure. This evening and yesterday the *aurora borealis* has been visible.

17*th March.*—The wind having shifted to the north-west, I decided on starting, notwithstanding the force of the wind and the drift we received on our backs. I remarked, with pleasure, that those we were leaving behind were not unmoved as they squeezed our hands; but, to dispose them by my example to cheerfulness and alacrity, I gave the signal for three hurrahs, in John Bull fashion. All things considered, it is they who are to be pitied for remaining behind. Captain Leask, Mr. Hepburn, the doctor, and the cook, now compose the crew of the little *Prince Albert*, and they will have a hard task to preserve themselves from the attacks of *ennui*. Only one man, gifted with perhaps too prominent a nose, complains of frostbites. Slept at the first encampment.

18*th March.*— We are wakened rather earlier than we should have liked. One of our dogs, pursued by the others, took refuge on the roof of our snow-house, broke it in, and fell down amongst us, bringing with him a considerable amount of snow. The poor animal looks so silly and amazed, that we repress our first impulse of annoyance, and, profiting by our early rising, we reach the third encampment the same day.

19th March.—At eleven o'clock this morning, I once more had the pleasure of surprising Mr. Kennedy by our prompt return. Foxes, no longer kept at a distance by the dogs, have paid our friends frequent visits; seven have been killed, as well as a white hare and two partridges, or willow-birds. Bears of every size have likewise been prowling about the neighbourhood. The following statement will show how tame the Arctic fox is:—One which had been caught alive, was adorned with a copper collar, bearing an inscription, and dismissed with as much noise as possible, amidst everything which might terrify him; and, in spite of all this, he returned the next day to the same place to take his chance of being killed. Several times, when the dogs did not see them, or were not chasing them, they seemed to take but little heed of the presence of man, running, leaping, hiding behind an iceberg or a heap of snow, and looking on which side the enemy was, only having recourse to flight when danger became too immiment. The Americans, during their wintering in the pack, could not get rid of two foxes they had on board, and which ended by dying in the hold; they are decidedly the most graceful and gentle animals in this region. Last Sunday Mr. Kennedy examined the road in a southern direction; it appears to be excellent, the ice looking smooth and accessible as far as the eye can reach from the tops of the hills.

23rd March.—We were certainly favoured, for our return has been followed by a continuation of bad weather, which would have hindered us considerably had it overtaken us on the road. The last few days have been spent in preparations for a final departure; such as repairing our sledges—two of which threaten to fall to pieces—making shoes, &c., and all the occupations of Arctic travellers, not very amusing, but at least new to me. We find comfort in the bad weather, which is annoying us, by hoping that it will perhaps take the place of the equinoctial gales, which we must expect. An oven now adds its heat to that of the two stoves, which we are sometimes obliged to keep in all night; two men are constantly employed baking necessary supplies for our journey.

24th March.—For the first time I have taken satisfactory observations. The snow-drift, even more than the cold, has hitherto been a constant impediment. I took these observations at the foot of the grave of Sir John Ross's carpenter. This grave is formed by a pile of stones heaped round the coffin, the frozen earth being so hard that the sailors had been unable to break it.

26th March.—This morning the refraction is so great, that we mistake a fox lying on the summit of a small rock near us for a bear, whose limbs and movements we each think we can discern through a telescope. We divide into two parties, each of which is to attack him on a different side; and armed, the one with guns or pistols, the other with harpoons or bayonets; when suddenly the animal scampers away with a swiftness of which the bear is incapable; and Mr. J. Smith, having started in pursuit, soon brings back an unfortunate fox, who pays dearly for our mistake. In the afternoon, a real bear is pursued as far as the moveable ice, upon which it is impossible to follow it.

I have omitted to mention that the sea-floe has been constantly separated from the land-floe by a stream of running water, the width of which varies with the force and direction of the wind. This stream does not appear to enter far south.

27th and 28th March.—Heavy westerly gale and snow-drift. Our preparations are ended, and to-morrow morning we shall set off, if the weather is favourable. Sunday is devoted to rest and prayer, as it has always been, unless circumstances have occurred to prevent it. It is not only on account of the moral reflex which religious observances naturally cast upon officers, but more especially because I feel impelled thereto by the inward voice of my conscience, that I like from time to time to take refuge in prayer.

A week ago, I accomplished my twenty-sixth year; in the last ten years I have passed through more dangers than men of my age usually meet with. I have passed safely through those trials; and when I speak of my lucky star, or of predestination, I do not mean that I place my confidence in anything astrological; that would be too absurd and too impious. No; my confidence is placed higher; I do not believe that Providence has guided and sustained me hitherto to abandon me in the midst of my greatest trial. I do not care to lose myself in the labyrinth of religious systems, in which I believe there is little besides sophisms, more or less fallacious; but I listen to that inner voice which tells me that we are not thrown upon this earth by chance, without compass to guide our conduct, without guardian to protect us. My prayer is offered up direct to the throne of the Almighty who created me, and renews my existence day by day.

Before undertaking a journey, the chances of which it is impossible to foresee, I will once again place myself in the midst of all those I love, and ask the blessing of Heaven upon them and upon me. Full of confidence in the Divine mercy, I acknowledge all my imperfections; and if my conscience is at rest, it is because I trust, not in my own justification, but in a goodness as inexhaustible as it is boundless.

And now, let the struggle with the physical and moral perplexities of life on earth come when they may, I feel full of strength, of courage, and of hope. My brother, my Alphonse, if my counsels cannot be given to you, remember, dear child, before beginning any arduous undertaking, always to invoke Him who has said, "Knock, and it shall be opened! Ask, and it shall be given!" And then, with thy conscience to guide thee, and thy heart in thy hand, march fearlessly on!

29th March.—At eight o'clock the thermometer stands at $-26°$; at twelve, $-12°$; at ten, $-24°$. At last, here we are on the high road to the unknown and unforeseen. At four o'clock in the morning, the weather having cleared up a little, we begin our packing, both of what we leave and of what we take with us; load our sledges, and put everything to rights in this dwelling, which has neither our regrets nor our thanks, and, yet, to which we owe some gratitude, even for its imperfect shelter. Besides, who knows if .... Mr. Kennedy has not though it advisable to follow out his first plan of sending provisions *en échelons* on the road, and our twelve men accompany us. Eight of them—it has not yet been settled which—will return from Brentford Bay to the ship, forming a fatigue escort; the carpenter and some of the men will then return to Batty's Bay and Fury Beach, to repair the strongest of the boats.

The plan of the journey is fixed thus:—We are to follow the coast as far as Brentford Bay, where we shall traverse the isthmus, five miles in width, which is supposed to exist between the two seas; perhaps there may be an uninterrupted passage. Thence we shall descend to the Magnetic Pole, and, according to the time of year at which we get there, to the state of our provisions, and to our meeting or not with natives, and their ability or inability to give us provisions for our objects of exchange, we shall either make that the limit of our journey, or proceed further south. Having left concealed provisions on the road in order to lighten our baggage, we shall return along the same coast; and, instead of again crossing Brentford

Bay, ascend northward to the limit of Sir James Ross's explorations, and cross Creswell Bay, in order to reach Somerset House. Mr. Kennedy not having adopted my plan, which was to visit the other coast first, I have directed my attention to this one; and I think that, since our leader does not adhere to the first and original plan of our expedition, his idea may have the following advantage over mine, namely:—That if Sir John Franklin has landed on any part of Boothia Felix, coming from the westward, we shall meet with his traces, or the Esquimaux give us some information; in the contrary case, the question will only be settled in the negative after exploring the eastern coast, less to discover traces of landing, than to see if any man is there who has survived the disasters of the *Erebus* and *Terror*. My only objection is this:—If we do not meet with any natives at the Magnetic Pole, we must go far south to find them; whereas on the other coast the two Ross's found them much further south; and I think it is essential to communicate with them as quickly as possible, in order to reassure the unfortunates who have, perhaps, long since given up all hope, or to obtain information of the greatest use to ourselves.

News is far from circulating with rapidity amongst these tribes, who move only in the narrow limits of fisheries or hunting-grounds; but we know from Sir James Ross that the Boothians communicate with the western natives, perhaps with those of Victoria or Wollaston Lands; from Dr. Rae we know them to be in relation with those of Repulse Bay; therefore it is not improbable that, if the whites have landed either there or in the neighbourhood, we should hear of it.

For my part, I attach the greatest importance to this inquiry amongst them, because it is impossible that a party of Europeans, however reduced in numbers, could pass unnoticed in their spring and summer explorings along the coast; whereas we may pass close to very distinct marks without perceiving them. It was thus the Americans passed by Cape Riley without seeing the traces which Captain Ommanney found there the next day. As to the plans and objections which I write down here unreservedly, I only venture to suggest them to Mr. Kennedy, who, doubtless, knows better than I do what should be done; more especially because I see that he takes offence very easily. My feeling is, that my opinion being asked by my commanding officer, my duty is to tell him all I think, and not to suppress any of the objections which, rightly or wrongly, present themselves to my mind; and then let his authority or greater experience decide.

Slight easterly wind; the land grows gradually lower, and is in many places so thickly covered with snow, that we only know it to be land by black spots peeping out here and there; the ice is perfectly smooth, and we perform from sixteen to eighteen geographical miles, with ease, between ten in the morning and seven at night. Our caravan is composed of fourteen persons. Mr. Kennedy goes first, in order to choose the best roads, and our four sledges, to two of which the dogs are harnessed, follow in joyful procession; our journey being enlivened by hope, even the unvarying monotony of the coast is not without its pleasure, because it is new to us. Mr. Kennedy has given me no order as to what I should do; but I willingly and cheerfully harnessed myself to one of our sledges, in order to set an example of helpfulness. I should blush to profit by my position, to be idle whilst others work, the more so because Mr. Kennedy wishes the instruments of observation, and the few indispensable works we carry with us (a Nautical Almanac, a table of logarithms, our pocket-books, an azimuth compass, and two small pocket compasses) to be placed with the rest of the baggage—an arrangement which might cause them to get out of order, I fear, if I did not remain close enough to watch over them myself.

The enormous snow-house required for fourteen people takes us nearly five hours to build, so that it is nearly one o'clock in the morning before we are all tucked in. Mr. Kennedy tells me the doubts which harass him. Will not our men be sickened with the expedition? It is, I think, wisest to set out with moderation, and to increase the distance when the habit of walking has been acquired; however, one day's fatigue will certainly not kill us. A mile north of our encampment, Mr. Kennedy, who followed the shore to examine it whilst we walked on the ice, which is easiest for our sledges, found two empty meat cases close to the bones of a white whale. These are, no doubt, the traces of Lieutenant Robinson's last encampment, and to-morrow we shall see the cairn he erected before returning to the north.

30*th March.*—At eight o'clock the thermometer stood at $-22°$; at noon, $-10°$; at nine, $-24°$. South-easterly breeze; cloudy sky. We only do twelve miles this day.

Whenever the coast or shore forms a little creek, we traverse it from point to point, whilst Mr. Kennedy remains on shore, on the look-out for something interesting. We have not seen Lieutenant Robinson's pyramid of stones, although fourteen pair of eyes (minus one) looked everywhere for it; the cross which formed its summit has been probably knocked down by the wind, and the pyramid demolished by the animals. If the Esquimaux had come so far, this would certainly have put them on the trace of Fury Beach.

Our present plan for our night lodgings

is to build two snow-houses with a contiguous wall, and make a door of communication. Encamped half a mile from shore on the ice.

31st March.—Thermometer, eight o'clock, − 26°; at noon, − 11°; at eight, − 15°: cloudy sky and easterly breeze. We are at the mouth of Creswell Bay; but, as the ice does not seem loosened from the further end of the bay, we make immediately for its southern point, Cape Garry. The ice is now very irregular, as it always is wherever there is a current during the season of running water; and at the end of twelve miles we encamp on the floe, and the faith of treaties, without our sleep being in any way disturbed by the fear of being carried out to sea.

Mr. Kennedy has almost a desire to cross to the west of this bay, which I should be very glad of, as we should then at once attain the last spot seen by Sir James Ross in 1849, and we should thus have an opportunity of regulating our chronometer; but, on the other hand, the isthmus is here twenty miles wide; the shore seems composed of lofty and irregular hills, among which our sledges might get broken; whereas at Brentford Bay it is not more than five miles wide: decidedly we shall go to Brentford Bay.

1st April.—Thermometer, at six o'clock, + 5°; at noon, + 7°; at eight, + 7°. Gentle westerly breeze; thick, cloudy weather; snow in the evening. After a march of thirteen miles, we find ourselves on the coast, at a distance, we suppose, of five or six miles west of Cape Garry; the land is very low, and when on shore we can see no point or projection deserving the name of a cape. We encamp at a distance of four miles from the lands most advanced to the east. Mr. Kennedy found on the coast the remains of some twenty Esquimaux huts, with the whales' bones always present in an Esquimau encampment, because in their peregrinations they stop wherever they find something to eat, and a stranded whale is a windfall, of which they leave nothing that can by any possibility be digested. Fresh traces of musk oxen, and their footmarks, since the last fall of snow, indicate that they frequent these shores in considerable numbers, either in their migration northward, or as an habitual residence. The shore, or rather the low flat lands, which seem to extend for a mile to the foot of the hills, are covered with grass and heaths. As the notes above indicate, the cloudy, snowy weather, quickly sends our thermometer up.

2nd April.—Thermometer, at six o'clock, + 1°; at noon, + 12°; at ten, + 13°. Heavy cloudy weather. We turn Cape Garry, and encamp in Fearnall Bay; the land is all low, as far as we can judge, while keeping in shore as closely as possible, yet remaining on the ice on account of the gravel and shingle, which would damage our sledges. We notice very fresh traces of a bear, and those of some snow-birds; the latter are looked upon as forerunners of warmer weather, which, strange as it may seem, is not at all pleasing to us, for to-day the work of hauling was made very unpleasant by the heat.

3rd April.—Thermometer, at seven o'clock, + 3°; at noon, + 4°; at eight, + 1°. Heavy, cloudy weather. At the south point of Fearnall Bay, we find tufts of hair and pieces of flesh which have belonged to a deer. At first we thought we had found one of the Esquimau depôts, or hiding-places; but traces of wolves, and snow sprinkled with blood, sufficiently indicate that the spot had been not long before the scene of one of those frequent struggles between the wolf and the reindeer.

The land still continues flat; in the afternoon a gale blows from the south-west, and about five o'clock compels us to encamp about two miles north of Mount Oliver.

4th April.—Thermometer, at eight o'clock, −18°; at noon, −1°; at six, −13°. The bad weather continues; strong south-westerly wind and drift in the morning; the weather clears in the afternoon. We are, perforce, condemned to repose; I say condemned, because the extreme lassitude which we felt for the first few days has greatly diminished, and will, doubtless, disappear with a little more use. We sleep like tops, and in virtue of the proverb, *qui dort, dine*, it has been resolved, in order to economise our provisions, that on our days of rest we should only make one meal. We also burn as much moss as possible, in order to save our firing. Those who have not travelled under the same circumstances, would perhaps laugh to hear our schemes, as economical cooks, and the saving we endeavour to make in our daily expenses; but this is of great importance to us: a few days' provisions, more or less, may endanger our existence, or have a decisive influence on our operations. Half the world, it is said, is ignorant of how the other half lives; but whether it interests others or not, it is of the greatest importance to us that the moss we find should be dry, instead of wetted by the snow, and should burn easily, in order to give us abundance—of what? of water—nothing more!

5th April.—Thermometer, at five o'clock, $-23°$; at noon, $+15°$; at eight, $-24°$. Gentle westerly wind; the atmosphere clear all day. Two miles south of Mount Oliver we fell in with the traces of some dozen Esquimaux huts, and their usual accompaniment of whales' bones. We march along rapidly, following the coast to the south. At noon, the height of the sun tells us that we are about a mile and a half north of the spot where Sir James took possession of these regions in 1829. We are equally at a loss with that officer to distinguish the islands from the small peninsulas attached to the continent, amidst the low and inundated lands which form the northern part of Brentford Bay. On one portion of this land, which Mr. Kennedy explores to westward, whilst we keep outside, we find footmarks, which we all believe to be those of two Esquimaux, a man and a child, on the friable snow—that best adapted to retain similar traces. Although these do not appear of recent date, we immediately conceive a hope of meeting with the natives; and here are our reasons:—The deer, in their migrations towards the north, or the sea-shore in summer, always travel in large numbers; and they are hunted all the more successfully because the country they have to traverse is composed of narrow valleys or passages, out of which they can with difficulty escape.*

* The land-marks, or stone-marks, which are so often met with in the places through which the deer are accustomed to pass, are explained to me as follows:—The deer migrate at certain times, and in certain directions. At these seasons the natives repair to all narrow passages, such as the isthmuses,

The narrow isthmus which unites Boothia to North Somerset is, therefore, in all probability, an excellent hunting-ground, which the natives must frequent at this time of the year. Mr. Kennedy found, on the lands he visited, traces in every direction of the reindeer, and their continual persecutors, the wolves.

To the north of the land which we suppose to be Brown's Island, there extends an inlet running to the westward. A thick mist rising from this inlet indicates the presence free water; but before we can ascertain this, night falls, and compels us to encamp on the the north shore of the island. The lands surrounding us are more lofty than those we have hitherto met with, and the islands seem to be rocks of granite; on the shores we find in abundance moss, which we can use as fuel. Hitherto we have only burned wood

valleys shut in by lofty hills, &c.; but, as these animals are timid and prudent, it is easy to deceive them by placing stones on each side of the road. Thus, the hunter stands at the further end of the valley; the reindeer, alarmed by the first stone, takes another direction; after pursuing that for some way, he meets with another stone, which also induces him to alter his course; and so on; and at last, after a series of tackings about, the poor beast falls into his enemy's power. Branches or anything else may take the the place of stones.

and coal, in order to reserve our spirits of wine, which are infinitely lighter to carry for those who go south; but one of the tin jars which contained it has been broken on one of the sledges, and part of the liquid lost. In the evening, a bear came to scent our baggage at only a few *mètres* from us, and during our supper-time the dogs gave him chase, and were soon out of sight.

6th April.—Thermometer, at seven o'clock, $-13°$; at noon, $-1°$; at eight, $-14°$. According to yesterday's thermometrical observations, the temperature varied 38° of Fahrenheit between five o'clock in the morning and twelve, and a little more between twelve and eight in the evening. If these extraordinary variations of temperature have a bad influence on the health, we do not perceive it; it is certain that in a milder climate we could not endure such variations with impunity.

Fine weather; gentle northerly breeze. Mr. Kennedy has determined on sending back our fatigue party, which has at least relieved us of the weight of all the provisions and fuel necessary for the journey from

Fury Beach to Brentford Bay, a seven days' march, making more than a hundred kilogrammes. Some provisions are to be sent from the ship to Somerset House. What remains to us, and forms our cargo, consists of about four hundred pounds of pemmican, a hundred and fifty pounds of biscuit, a small quantity of tea and sugar, and five gallons and a half of spirits of wine. Our spirits of wine weighed about thirty-eight pounds; the whole making nearly six hundred pounds' weight of provisions. Besides this, we have each a blanket, a buffalo-robe between two, a macintosh cloth, the sextant, the air-boat, the things for barter with the Esquimaux, three guns and their ammunition, and twenty pair of mocassins: we carry upon us all the body linen we are to use during the journey. The load of a dog in Hudson's Bay, for a journey of some length, is estimated at a hundred pounds; our five dogs were, therefore, just sufficient to drag our provisions, and what remained for us five to carry must have weighed at least as much.

We are now only six: Mr. Kennedy and

myself, Messrs. J. Smith, W. Adamson, A. Irvine, and R. Webb, the ex-sapper. These are men on whom we can rely. Mr. J. Smith is the one who best understands constructing a snow-house; another is our dog-driver, and moreover a good shot—a thing not to be disdained, as we may have to depend on our hunting.

Mr. Kennedy has not yet told me what he intended doing (about taking observations) before the departure of the ship. Here is the plan I have determined on for myself, if I am permitted to put it in practice:—The chronometer was allowed to stop whilst at Fury Beach, and I could only take one series of observations; perhaps we shall not be able to reckon on its regularity, and I intend to find the longitudes as well as can be done by lunar distances. Besides, if the coast joins Cape Bird at the Magnetic Pole, it lies almost entirely on a meridian. In any case, not having been able to regulate the chronometer at the spot on which possession was taken, I intend doing so at Cape Bird, by the Magnetic Pole, going and returning by the point most distant from the places visited by Sir James Ross, and then again at Fury Beach. As to the latitudes, I shall look for them, as often as possible, by the excellent method of taking two altitudes at a short interval; and the astronomical bearings which I shall take at the same time, with observations of the horary angle, will give me a means of ascertaining the longitudes. I intend to determine the variations of the compass by every possible means, but using the simple method of the first vertical for general use on the road. Unfortunately, Mr. Kennedy has been told, what is doubtless true, but has been falsely applied, that we are not here to make scientific observations, but researches. With this idea, it is to be feared that the simple observations necessary to determine our position, and show how and where our researches were made, may be mistaken for scientific observations: it is not enough to walk a great deal and to a great distance.

Two of our men remain at the encampment to dry the blankets, and Mr. J. Smith and myself go to the other side of the island; whilst Messrs. Kennedy and Adamson go westward, in the direction of the free water seen yesterday, in order to reconnoitre the ground, and look for a passage to the western sea. After following the indentations of the coast, and trying to discover the loftiest hill, we clambered up the granite rocks with some difficulty, and, having reached an elevation of a hundred feet at least, discovered the bay to the south and west of us.

Having come with the preconceived and deeply cherished notion of finding a passage, my first impulse was an exclamation of joy at the sight of the sea; but the sun shining directly above it showed that it lay to the south, and not to the west; for it was almost mid-day (we had left the camp at ten o'clock), and consequently the sun was in the south. I had given my little pocket compass to the party returning to the ship, by Mr. Kennedy's orders, and had none with me, Mr. Kennedy having set out with Mr. Adamson before Mr. Smith and myself, and having the habit of carrying with him our only azimuth compass and his own pocket compass.

I had with me a book of notes taken during the winter, with a description of those portions of the coast we might possibly visit, and a copy of Sir John Ross's map of Brentford Bay.

The land to the south appears very flat, either because the sun dazzles us, or perhaps on account of the vapours rising from the snow; but we connect it in an unbroken line with the western land, upon which we can point out the various inlets and rivers indicated by Sir John Ross; only that these lands, from which, according to our calculations, we should only be seven or eight miles distant, seem twice as far off. We cannot see northwards because we are not on the highest parts of the island.

We keep on to westward after descending to the sea, which we mistook for a lake; and, after once more ascending the steep rocks covered with melted snow, without making any fresh discoveries, return to the encampment.

On our journey homeward, we found an irregular strait between the island on which we had encamped and the land further west, which is probably Ross's Long Island.

We saw some ptarmigans, and numerous deer traces, which Mr. Smith declares to belong to last year. In the morning, and on our return, we find the snow, with which the ice is covered, all over bear-tracks of various sizes; amongst others, those of a she-bear and two young ones (Scoresby says they bring them forth as twins), which appear to have capered round her. On following their tracks, we arrive at a hole between two large icebergs, which had been probably enlarged and improved by the female before she littered. As we have but one gun, we do not care to disturb this interesting family; and, without ascertaining whether they are in their burrow, we leave their habitation in peace. Sme distance off we obtain proofs that, for once at least, this she bear has eaten grass. It is not known whether they suck their paws by way of sole nourishment during the winter, like the rest of the *plantigrada;* but is grass an habitual diet, or simply taken as medicine? Mr. Kennedy, on his return from his excursion, says that the land in a westward direction is some thirty or forty miles off, and on the other side of the sea, which he has also seen in the west. These are, I think, the islands seen by Sir James Ross. They kept alongside the open water (our yesterday's conjectures had not deceived us), in which he saw ice-blocks running five knots an hour: unfortunately, this water did not extend far. They likewise saw numerous deer-tracks, and during their absence a doe and her fawn crossed the route they had taken in the morning.

7th April.—Thermometer, six A.M., − 25°; noon, − 16°; six P.M., − 32°. With our two sledges and the dogs we set out on the route taken by Mr. Kennedy yesterday, and we find on our way, besides the principal piece of water, two smaller ones, over which, as Mr. Kennedy observed yesterday, the ice-blocks move with great speed and rapid gyrations. The air in the neighbourhood of these pieces of water is loaded with chilling moisture, and the wind, which follows all the outlines of this coast, blows in our faces, and makes us feel the cold more keenly than we have done for a long time. In the evening almost all of us have our faces spotted with frostbites. In my case, my poor nose suffers most—enough even to make me rather uneasy, for I have already a scar which divides one of my nostrils in two.

At noon we reach an inlet perpendicular to the general direction we have hitherto followed. I put my feet into a heap of snow, or melting ice; but, having carefully scraped my mocassins, the remaining moisture froze very quickly, and my feet were cold, but not wet. It is necessary, however, to pay attention in such cases, for foot-gear that is not supple may cause the toes to freeze.

Going southward along the inlet, the general direction of which is from N.N.E. to S.S.W., we come upon several tracks of foxes and bears. At six we encamp at the southern extremity of the western margin. An observation of the setting sun gives for the variation of the compass 144° W., whilst at Fury Beach it was only 132°. At the entrance of that bay, the Rosses, in 1829, found the dip of the needle to be 89°, whence they rightly concluded that the Magnetic Pole was not far off. The recollections of the interesting observations made on this subject made me often regret that during our long winter we have not had the necessary instruments to repeat them.*

8th April.—Thermometer, eight A.M., − 16°; noon, − 4°; eight P.M., − 33°. Light breeze from the north. We make for the north-west part of the bay, where we think we see an opening; but, after toiling for twelve miles (to N.W. ¼ W) over very uneven ice, we descry land at the bottom of the opening, and so find it closed before us. Captain Kennedy attributes the rough state of the ice to a current which exists in a passage to the sea on the west; but why should it not be a current all round the bay? I am not fond of launching out into conjectures in geographical matters, and I believe only what I see; though, after all, his hypothesis is possible; only hitherto we can derive nothing. A halo.

* It is extremely difficult, not to say impossible, to reconcile Bellot's account of the events of the 7th, 8th, 9th, 10th, 12th, 16th, 18th, and 20th April, 24th May, 9th June, 1852, &c., with that of Captain Kennedy (7th April), in his "Short Narrative of the Second Voyage of the *Prince Albert* in Search of Sir John Franklin." London, 1853.—*Note of M. de la Roquette.*

9th April.—Thermometer, seven A.M., − 20°; noon, − 6°; eight P.M., 34°. We now shape our course a little more southerly (W.N.W.), and towards another opening. The land appears continuous to the north. Captain Kennedy and two others of our party are beginning to be affected by snow blindness. Medical writers differ in opinion as to the causes of this disorder; some attribute it to the powerful reflection of light from a white surface; but it is certain *(experto crede Roberto)* that the effect produced is the same in the thick foggy weather so common in these regions. The preservative means employed are a piece of black or green gauze or crape, or very fine wire goggles, seen through which all objects assume shades less offensive to the eye; but I think they should only be used when the light is too glaring, and irritates the retina. Our travellers used them at all times; but it strikes me that, when the eye can hardly distinguish the objects before it, to interpose a screen which renders those objects still less distinct, can only add to the causes of inflammation, by occasioning a greater and more continuous tension of the visual organs. Until I have proof to the contrary, therefore, I shall use these preservatives only when the sun shines with unusual lustre.

Breeze from the east. After five miles we halt, at noon, on a sort of reef, near which the latitude observed is 71° 58′; and we march another six miles over the same sort of ice, which makes our progress very slow, for the intervals between the blocks are filled with soft snow, into which dogs and men sometimes sink more than a foot deep. At sunset we see land once more, at the bottom of the opening towards which we are marching. We resolved to persist in the same direction, otherwise we might range the whole outline of the bay without finding the passage we are in search of; and which after all is not indispensable to us, since the isthmus here is but five miles wide, and even less. I observe, too, that we are much more to the west than the coast marked on the chart at the bottom of the bay. Encamped on the ice.

10*th April.* — Thermometer, seven A.M., − 8°; noon, − 8°; six P.M., − 6°. Strong breeze from the east, which increases continually, and raises in great drifts the snow covering the ice, which is now much smoother than the two preceding days. We halt at noon to take the latitude (72° 3′) in a very disagreeable mist, which hides the ground over which we are walking, and which we imagined to be still far off; when, to our great surprise, and after only a few minutes walking, we find ourselves on the beach. We bear away a little to the south, to enter the inlet we had in sight yesterday; but the breeze and the drift increase to so inconvenient a degree that we are forced to halt.

We proceed immediately to the construction of our snow-house, quitting the work from time to time to stamp and warm our feet and hands. In two hours we had built our walls, two of us acting as engineers, while the rest prepared materials; one cutting or sawing the snow into long rectangular pieces, another taking them away on a shovel or a cutlass, and two others carrying them to our masons. What occupied us most time in the beginning was the building of the roof; but, setting aside the rules of art, and considering that our dwellings are but temporary, we lay the pieces of snow transversely, as well as we can, on the top of the walls, so as to meet above. An Esquimau would smile, perhaps, at the sight of our *chefs d'œuvre;* but, as they shelter us sufficiently, that is all we can desire. Besides, celerity is, above all, our aim; for nothing can be less agreeable than dabbling in the rubbish of this sort of masonry in a temperature of 20° or 30° below zero, especially if one has also to endure the wind and the infernal drift. The building finished, we all set to work to cement the joints with snow; and when the drift has, moreover, covered the hut with an additional layer, there is no escape for the heat produced within by our spirit-of-wine kitchen, and by our breath. A European would hardly believe that we were often incommoded by heat. It is necessary, however, that all the apertures should be well closed; otherwise a little hole, no bigger in diameter than the barrel of a quill, often becomes a funnel for the admission of a bushel of snow when it blows or drifts.

All toil brings its own reward. We are

very tired, certainly, when, after a day's march, we must apply ourselves, in the first place, to the still more fatiguing work of building, and of carrying or sawing the hard snow, which is as heavy as cut stone; but no one can imagine the sense of pleasure and comfort we experience when we are at last able to shut our door, and stretch ourselves out on our sacks. Our poor dogs themselves, after having gluttonously bolted their ration of pemmican, which they digest, no doubt, like the boa, throw themselves carelessly on the snow, never heeding that which falls and covers them completely, or rather regarding it as a supplement to their thick fur. Happy dogs! who have no building to do, and are perfectly indifferent as to the route we take, provided it is one that does not entail upon them too much flogging.

11*th April.*—The snow-storm growls outside, so that we dare not venture out of our hut, lucky as we are to enjoy its warm and comfortable shelter; but we are not without uneasiness for those who left us on Tuesday morning. We are afraid they are still about the middle of Cresswell Bay, on the ice of which this storm may have some effect. We cannot but implore for them that same protection of which we so visibly experience the effects. How can we refrain from admiring the Providence which changes into a saving shelter that snow which would very quickly be the instrument of our destruction! Do we not enjoy real welfare, whilst all around seems conspiring for our ruin? What strength is inspired by confidence in Him, without whose permission not a hair of our heads can fall!

12*th April.*—Thermometer, ten A.M., $+7°$; noon, $+10°$; ten P.M., $-7°$. In the morning the bad weather seems disposed to continue; but we take advantage of a respite to set out a little before noon, and we are favoured during the rest of the day by a light breeze from the east. We traverse the little bay we entered on Saturday, the 10th. After a course of about three miles to the west, and after ascending a very gentle slope, we reached ground almost uniformly flat: a hillock a little to the south is the only object that breaks the level. Then changing our direction, we marched about five or six miles westward, over soft and dry snow, which we trod with difficulty. Weather snowy and cloudy. We cannot see very far, but we advance with pleasure, reckoning that to-morrow we shall see the coast of that western sea towards which we have been plodding longer than we had expected, without coming upon it.

13th April.—Thermometer, seven A.M., $+4°$; noon, $+6°$; eight P.M., $+2°$. It seems that everything is against us; we are in the midst of a fog so thick, that it is impossible to keep the same course for many minutes together. To make matters worse, the needle has become exceedingly sluggish, and the compass often indicates three or four different directions as the one we want to take. *The compass sleeps*, as we say in nautical language; and it is only by shaking it, and comparing it with our small pocket compasses, that we can have any confidence. Perhaps the coast is not more than a few hundred yards from us, and we are marching parallel with it. Those among us whose eyes are in a bad state, suffer much from the efforts they are obliged to make even to look at their feet. In fine, it did not appear that we were advancing to the end of our journey by so uncertain a march, and at two in the afternoon we encamped.

Yesterday and to-day every elevation of ground was hailed by us as being perhaps one of the ridges, the further slope of which forms the coast. The blue clouds seemed to us to announce the presence of the sea, and even the reflection of flowing water. We remark this evening, as we have done on many previous occasions, that the snow, which lies on the ground in not very thick layers, consists of a sort of crystallisation, due, no doubt, to filtration, or to the precipitation of aqueous vapour, caused by the warmth of the ground or the herbs; and that these pieces of snow are, to an extraordinary degree, transparent, and conductors of sound.

The hillock seen on Saturday, the 10th, is now three or four miles to the east of our encampment, and not far from its foot runs a small ravine, which no doubt abuts on the beach.

Though the thermometer is not much above zero Fahrenheit, we find that the warmth increases very sensibly with the cloudy state of the sky—a fact for which the laws of physics perfectly account.

14th, 15th, and 16th April.—Continuation of the same weather: the fog is so thick that we cannot venture a few steps from our house without quite losing sight of it; and the compulsory state of inactivity to which we are thus reduced, is certainly more intolerable to us than the most rugged route we have yet traversed. It increases our vexation at seeing our provisions consumed without our making any progress; moreover, the most stoical philosophy might be put to the proof in a snow-hut eight or ten feet long, five or six wide, and three or four high; for the heat caused by our long residence has damaged our masonry. On Friday, during a short interval of clearer weather, Captain Kennedy went a little a-head with two men, to reconnoitre the ground, and get sight of the sea, which we are impatient to behold; but he did not succeed, the land, as far as he could see, being almost level.

17th April.—Decidedly our impatience does not mend matters, and there is nothing for it but resignation. We take to the pipe as a solace for our disappointment, and smoke with such persistence and regularity, that we have no need of a watch to tell the time of day. "What o'clock is it, Dickie?" "So many pipes, sir." With the ground in a tolerably good state, and finding the coast as it is marked on the chart, we should now be at the Magnetic Pole—perhaps among the Esquimaux!

18*th April.* — Thermometer, six A.M., $-4°$; noon, $-10°$; six P.M., $0°$. To our great delight—for we are tired of rest— the weather at last allows us to resume our journey under a radiant sun and a light breeze. Close to our camp we see a grey wolf, whose tracks show that he must have been prowling about near us during the night. If the rascal is hungry now, he has not always fasted, as appears from certain indications containing bones of deer. He follows us the greater part of the day; keeps a respectful distance, after receiving a hint from a ball to that effect, but he pursues us with his howlings. It is in this way that in the Hudson Bay territory they often try to decoy the dogs, with which, in these regions especially, they have some kindred. Nothing can be more lugubrious than the plaintive cry that still resounds in our ears after the animal has disappeared. Whether it was from heedlessness or their sense of security, our dogs did not seem to pay the least attention to him.

The snow is beginning to feel the effects of spring, which increases the difficulty of our march over the undulating ground, broken with ravines, which we meet with to-day. It is clothed in many places with little tufts of heather and branches of the dwarf willow. The partridges, which are fond of the buds of this shrub, have dug out the snow with their bills; and we also see numerous tracks of foxes running south, or towards the spots frequented by the partridges. Thus the snow enables us to perceive, almost everywhere in these astonishing regions, that animals seem generally to proceed in groups of two species, one of which preys upon the other.

It has been so warm to-day as to allow of our enjoying a partial ablution with a few handfuls of snow. We have gained about twelve miles to the west, and a few to the north, and we encamp some miles from a pretty high hill. The observation at noon shows the latitude to be 72° 03″—nearly the same as that of Cape Bird, which, however, we have not yet seen. At noon we were near a ravine running north and south, the steep sides of which afforded us some sport, as we had to unharness our dogs and launch our sledges, which slided over the snow, and ran to a great distance, as a ship does when it is let go from the stocks. During the greater part of the day the sky to the west was loaded with blackish vapours, which we are pleased to regard as the blink of open water. I am not aware that any hygrometrical observations have yet been made here. The skin becomes so dry under the very peculiar effect of intense cold, that it is extremely irritable, and chaps as much in the sun as in the most piercing northeaster. I imagine that to this cause are owing most of the ophthalmias comprised under the generic name of snow-blindness.

19*th April.* — Thermometer, nine A.M., + 22°; noon, + 26°; ten P.M., + 22°.

Light breeze from the south; cloudy and snowy weather. We again make for the west, or endeavour to do so. About four miles from our camp we come upon a river or lake lying north and south, rather narrow, but appearing to stretch far away northward. In order to economise our fuel, by drinking before we are thirsty, we dig four feet down into the ice, and try, without success, to break the solid layer beneath by firing some balls against it. Our toil will not have been quite lost, however, for we fill our boiler and the rest of our utensils with fragments of this fresh-water ice, which will cost us less time and spirits of wine than if we had to melt snow. I was greatly surprised to see the bullets rebound from the bottom of the hole, though the pistol was pointed vertically. In September last I saw balls fly back after striking the ice, contrary to the laws of reflection.

We have every moment to consult the compass; and as the region we are now traversing is rather irregular, in consequence of the detour we make to avoid descents too steep for ourselves, or too stony for the sledges, we make little real way. After seven or eight miles of continual ascents and descents, the valleys generally running north and south, we come to a deeper ravine than any we have yet encountered. The ground gradually descends to a level so perfect and so unbroken by stones, that we begin to think that we have reached the sea without knowing it; nor was it until we had dug through the snow, and found the earth not more than a foot beneath it, that we were undeceived. The land we leave to the west, and which we see terminating in this sort of plateau, also contributes greatly to create or maintain this illusion. The experience of the few following days showed us the danger to a traveller of such preconceived opinions; for, expecting to find here the ice of the Polar Sea, the temptation would certainly have been very strong to believe in its existence.

In order to make up for the delays caused by the fogs from the 14th to the 17th, and urged by a very natural impatience, we pushed on towards the west, not halting till we had marched for twenty-two hours, during which we must, at the most moderate computation, have got over from fifteen to eighteen miles—that is to say, a degree of longitude to the west. Captain Kennedy has resolved that for the future we shall travel by night, and sleep by day. Our dogs suffered much from the heat, by which we too were incommoded during the laborious work of haulage : one of them even had an attack of epilepsy. Our new system of night travelling is especially favourable to the eyes, which are much less fatigued by the lustreless whiteness of the snow at that period. The valleys we have crossed generally contain water. The snow is also very soft, which increases our toil; but we cannot complain of this, for it spares our sledges, which would be soon knocked to pieces on a harder surface. We cross many tracks of foxes, partridges, and white mice.

20th and 21st April.—Thermometer, nine P.M., $-7°$; midnight, $-10°$; nine A.M., $-10°$. Light breeze from the south. We have travelled twenty geographical miles to the west over these snow steppes, which still present the same flat monotonous uniformity, never varied but by a stone at long intervals, or by a slight undulation, on which is found the favourite lichen of the reindeer. We still see numerous tracks of foxes, and several small burrows made in the snow by the Arctic mouse. It seems to us absolutely impossible to reach the west coast, which lies before us, unless the compass deceives us; but little as we trust its indications, on account of the vicinity of the Magnetic Pole, the distance we have traversed westward, which I have always computed at the lowest rate, places us at more than four degrees from Brown's Islands. The sun, during the intervals we have seen it, has always served us as a check upon the needle. In any case, it is evident that the sea, to the west of Somerset Land, is closed to the north of Cape Bird; and that the best way is to make straight for the north, where we must soon or late come upon the coast, and then march along it. It is very probable that the ravines we have lately crossed abut upon the sea; and if we do not find it to-morrow, after some ten miles march to the north, my advice would be that we should bend to the north-east. That direction, from our present latitude, even supposing it to be more than a hundred miles to the west (as Captain Kennedy thinks), would infallibly bring us to the coast on the following day.

Hitherto I hoped that Sir James Ross was right in his conjectures, but there can be no doubt now that he was mistaken; besides, he is not very positive in what he says as to the place where he was—that is to say, —— miles* north of Cape Bird. He examined the country; the clearness of the day, he says, was remarkably favourable; and the whole land, being very elevated, could be seen to a distance of a hundred miles. Now, the lands we have passed are so flat that we often took them for the sea itself, and could only give up that idea when we had ocular proofs to the contrary. It would not be surprising, then, if even at a less distance he made the same mistake as we did with regard to what was under our feet. The islands he marks on the map a little to the north of us would, in that case, be the several hills we have seen, or highlands still more north, which we could not distinguish.

* Blank in Bellot's MS.

For the present, we cannot reckon on the chronometer; and it is not until we are on board that we shall be able to say exactly where we have been.

Three alternatives were proposed to us yesterday by Captain Kennedy: — 1. To continue a western course for one day more. 2. To retrace our steps to Brentford Bay, where he expects to kill a great number of deer in a week, and with that supply of provisions to proceed overland to the Magnetic Pole, or to Cape Elizabeth. 3. To make for Cape Walker. My advice was to continue to march one day longer to the west, and afterwards to come to a new decision, according to circumstances. As for returning to Brentford Bay, and starting thence for the south, I should exceedingly desire it if it were possible; but I doubt much that we should be so successful in deer-hunting as Captain Kennedy supposes; indeed, he admits that possibility. If, after a week's hunting on the islands of Brentford Bay, we should not have been sufficiently successful, we should be forced by want of provisons to return to Fury Beach; and how disagreeable that would be! Otherwise, I should have been glad to go in that direction, for it is the one to which we have been sent by Lady Franklin. To do so, was part of the plan I proposed to him in February; besides, that point is more important to explore than Cape Walker, which must have been visited by one of Commodore Anson's vessels, or by Sir John Ross. We finally adopted the first of the three alternatives; and, as Captain Kennedy asks us all for our opinion, I took the opportunity this evening fully to explain mine, which would be, move north-east, in the hope of meeting the sea as soon as possible. Captain Kennedy desires to get due north, that we may remain on land, and on snow, which is less dangerous to our sledges than the ice we may meet with on the coast, and not in the interior. To this I reply, that if Sir John Franklin's expedition penetrated to the bottom of this *cul de sac*, we shall find traces of its passage on the coast, and not in the interior. "That is true," said Captain Kennedy; "but for the present I wish to get to Cape Walker as soon as possible, and the coast will be examined by the boats during the summer." Nothing remained for me but to bow, after having clearly expressed my own views.

21st and 22nd April.—Light breeze from the south; hazy weather. We steer northwards, at least as well as we can, for we have every moment to consult the compass. The only way in which we can direct our course is by taking for guide an object in the north, and marching towards it until we see another object in line with it, and situated further off; but after having stopped to choose such marks, and chosen a block which seems pretty big and distant, we come, after a few minutes' march, upon a small stone, the apparent bulk of which had only been magnified by refraction. Nothing can be more vexatious than these incessant impediments to our advance, or, at the same time, more fatiguing, especially for the eyes of the foremost man, who peers through the fog, looking out for these marks. From time to time, one of us takes the place of Captain Kennedy, who is usually a-head of our sledges; and, though we are all very attentive to follow the movements of our guide, yet every time the compass is consulted we find we have insensibly deviated from our intended route. Hence, though we have travelled nearly eighteen geographical miles, I do not think we have advanced northward more than eight or ten miles.

Three miles from our camp we came upon ground as flat as the rest, but destitute of those lichens which were so abundant before, and strewed with flat and rounded pebbles, which gave it the appearance of a beach; and we looked for a complete change in the character of the ground; but, after twelve miles marching, we came upon the same sort of plateau, covered with lichens, as we had seen on the 20th. On the morning of the 22nd the haze is so thick that, accustomed as we are to travel by night, it would be impossible for us to say, from the evidence of our senses, whether it is day or night at this moment. Saw a ptarmigan.

22nd and 23rd April.—Thermometer, nine P.M., + 12°; midnight, + 25°; eleven A.M., + 25°. Light breeze from the south-east and south; snow; thick fog. It is with the greatest difficulty we direct our course towards the north; we do not see even a stone by which we can guide ourselves. After having set up poles in the direction given by the compass, two of us walk in a straight line one behind the other, whilst the sledges follow, keeping the same line; or else one man walks in front and the other behind with the sledges between them. In spite of our care, the compass shows us that, after some minutes' march, we are sometimes 45° or even 90° from the north. The compass is very sluggish to-day. We think we have walked fifteen miles, but we can hardly have advanced more than eight or ten in the required direction. We have had no observations for some days, so that we cannot tell precisely the amount of our daily advance. We can distinguish nothing a hundred paces before us; and often our guide, who is thirty yards a-head, is half-concealed by the fog. Everything, then, contributes to make our daily marches long and wearisome. The ground, as far as we can judge, seems always the same.

23rd and 24th April.—Light breeze from the east; snow at the beginning of the night;

towards midnight the weather clears up, and promises us a fine day. Amidst these immense wastes, the various spectacles of the Arctic regions make a very peculiar impression on the mind. In this hazy weather the sun rises and sets without any of the splendour accompanying it on lighter days; a pale and enfeebled disk descends without pomp behind the whitish horison; the scene has a character at once sweet and sad, but not without solemnity.

We calculate that we have travelled fifteen miles; but, for the reasons stated yesterday, I do not think we can count on having advanced more than eight or ten miles north. We often catch ourselves tasting the snow, to satisfy ourselves that it is not brackish, so deceitful are the appearances of the region. Before noon the sky is more overcast than ever, and leaves us in the same ignorance of our position. We have often heard the cry of the ptarmigans or rockers near us, and have seen seven of them to-day, but cannot spare time to pursue them. One only, which remained within gunshot, has been killed. After plucking it, we let it freeze; and, doubting what our people of the north-west told me, I wish to ascertain for myself if this kind of game is really eatable raw and frozen. To my great surprise, I found it delicious.

For these three days I have worn snow-shoes, or *raquettes*. They suit me very well, and cause me none of the usual inconveniences described by Hood, except a great heaviness in the legs, which will, no doubt, soon wear off with use. The rest of our party has begun to do the same to-day; and this is of the more help to us, as these snow-shoes do not impede us in the work of managing the sledges.

24th and 25th April.—Thermometer, nine P.M., +22°; noon, +37°. The same thick haze; cloudy weather. Below a ridge of ground, on which we encamped, we find ourselves, almost without transition, on the ice, which appears to us all to have a pretty strong salt taste; but, after our late mistakes, we do not venture to decide in the affirmative. The marks left by the tides do not exist; nevertheless two long crevices may stand instead of them in this little creek, as it seems to be situated at the bottom of a bay, which we cannot distinguish. The coast on both sides, as far as we can see—that is to say, for half a mile right and left—lies north-west and south-east. This creek is closed to the south-west, and evidently belongs to the western portion of the lands seen by Sir James Ross. We continue our route to the north, and the ground becomes a little more undulating. To the east there appear to be islands, either very low or very distant from us. The sky having cleared in the morning, we can shape a good course. After a march of eighteen miles, according to our approximate reckoning, we find quite close to our encampment a bluish ice, as to the nature of which there can be no mistake. The sea cannot be far off. We saw a partridge, two snow-birds, and tracks of foxes.

In the absence of the blocks, or rough ice left by the tides, what seems to us a sufficient reason for believing in the presence of the sea, is the appearance of the small stones, which are now of a flat form set on edge, and split into thin laminæ, as is usual near coasts; whilst the stones we have hitherto seen have been round and smooth.

At last we get observations, which give us for latitude 73° 50′ 30″ north. This agrees pretty well with my calculation. For the first time for more than seven months the thermometer has risen above zero centigrade, and we inaugurate the milder season by a bold measure—sleeping in the open air without any other shelter than a little wall of snow between the wind and us.

25th and 26th April.—Thermometer, seven P.M., + 13°; midnight, + 20°; noon, + 22°. Light breeze from the south-east and east; mist thicker than ever. We try all possible means to keep in one direction, marching in single file, and guiding ourselves by the breeze and by a light flag; still, as on the preceding days, we find, on consulting the compass, that we are deviating twenty, forty —sometimes even ninety—degrees from our route. We are all snow-blinded now, and Captain Kennedy and I are obliged to relieve each other from time to time at the toilsome work of beating the track. The compass is still very sluggish. The ground has evidently assumed a new character, and appears to rise gradually from hill to hill.

We come again and again upon the quite recent tracks of four musk oxen, which we should, perhaps, discover not far from us, if the weather was favourable. The nature of the ground is no longer so advantageous for our sledges; and, if the mists continue, it will be impossible for us to advance rapidly It seems, then, more expedient for us to remain on the coast, the direction of which we may follow, whatever it be; so, after marching ten miles north, we bend off to the east for two hours before encamping at the foot of three rather remarkable hills, which are probably what we took yesterday for islands, or at least for the islets situated most to the west.

Whilst I was taking observations at noon, the wind threw down the spirit level of our artificial horizon, and broke it on a box of pemmican. Very fortunately for me, it was Captain Kennedy himself who placed it there. He would not bring our quicksilver horizon with us, as I advised him, because of its weight; and now we have none, and can take no more observations until we have the horizon of the sea. This is the more annoying as we might to-day, for the first time, have got lunar distances, which would give us the longitude, and enable us to do without the chronometer.

26th and 27th April.—Thermometer, eight P.M., − 13°; midnight, − 10°; noon, − 8°. We have marched of late through a sea of fog, through which the tops of the hills appear like so many little islands; and we now find that the islets seen the other day are nothing else. Strong breeze from the northwest. We pass a small lake at the foot of these hills. We move eastward; and, after ascending a pretty steep declivity, we come upon a stony level, where the wind has full play upon us. The thermometer has fallen 30° below the average of the last two days. Saw two more tracks of musk oxen going north, like those of yesterday. After marching for some time along the winding bank of a stream, now frozen, we encamp on its bed. It runs from west to east, but the violence of the wind hinders us from determining its longitude. We have travelled twenty miles in fourteen hours, but perhaps not more than ten miles in longitude to the east.

27th and 28th April. — Thermometer,

eight P.M., − 13°; midnight, − 5°; noon, − 4°. Clear weather; north wind. After three miles to the east, we come to the top of a hill, at the foot of which is a plain of snow, which we mistook at first for the sea. To the east, rather high lands, which we thought were islands. Discouraged, at last, we resolved to make no more conjectures, and, like St. Thomas, to believe nothing but what we touched with our fingers. We passed within five hundred yards of two reindeer, one of which appeared to watch whilst the other rested. Messrs. Kennedy and John Smith crept along the snow after them, in the Indian manner, placing their guns crosswise over their heads, to represent the horns of two other reindeer; whilst the men left with the sledges held up their arms, and shifted from one foot to the other. But whether it was that the imitation was too imperfect, or that the reindeer were too wary, they moved off out of gunshot. Here was a confirmation of my doubts as to the probable result of our hunting at Brentford Bay: I recollected how unlucky we had been with respect to all kinds of game.

At last, after four miles more to the east, or seven from the encampment, we came to the edge of a rather deep bay, the southern point of which is formed by the land seen this morning; there are several islands at its entrance. The most sceptical cannot doubt now, and we travel joyfully northward towards what appears to be the other point of the bay. No tide-marks. Two miles from where we are on the ice, the bay appears to be four miles wide, and to enlarge as we advance north. The breeze is still too strong to enable me to use a phial of liniment with which the doctor has furnished us, and with which I thought of making an artificial horizon—an easy experiment, on account of the black colour of the liniment and the camphorated spirit, which hinders it from freezing.

28th and 29th April.—Thermometer, eight P.M., $-13°$; midnight, $-3°$; noon, $+10°$. Clear weather. After some miles, we reach the foot of the cape, to the north of the bay. It is formed of what Mr. Kennedy thinks is sandstone, of a reddish colour, fragments of which cover the ground. We advance rapidly over the low beach, which stretches nearly eastward, whilst Captain Kennedy ascends one of the rocks near the cape, to have a view of the sea and the islands to the east. What we yesterday took for islands forms a continuous line, being, as he thinks, the west coast of North Somerset. For my part, since these lands have a north-westerly direction, and since the coast of Somerset runs northward, I think he is mistaken; but I think it better to say nothing, and allege my inability to pronounce an opinion in consequence of the state of my eyes. After passing this cape, the land sinks almost suddenly, not to rise again till much further north. We encamp on the coast, having travelled twenty miles to N.N.W. Towards the end of the night the wind shifts from north to east, and the weather becomes snowy.

Perhaps I was wanting in firmness in not opposing the declaration, that the land seen to the east is that of North Somerset (which the following days disproved); but I must let the captain bear the responsibility for his declarations. After passing the cape I have mentioned, Captain Kennedy turns to us, and says that it shall henceforth be called Cape Bellot; that the inlet shall bear the name of Mr. Grinnell, and the land that of Prince Albert. I beg leave, with many thanks, to decline this honour, and to unite with our people in bestowing it on our commander, to whom, I think, it is due before me. Absolute refusal on his part.

All the glory I can acquire here consists in having joined this expedition. In France it will be thought, perhaps, that I found here routes all ready, and as easy as possible; only the officers and men of the ship would be competent to bear testimony; but, God forbid I should ever appeal to them! After having given my voluntary services, I render myself this impartial justice, because I derive it not from my passions, but from the bottom of my conscience, that I have done more than could have been expected of me, considering my want of experience in these regions. Who will ever know what hopes, joys, and blessings I have sacrificed to undertake this cruise?

29th and 30th April.—Thermometer, eight

P.M., − 22°; midnight, − 20°; noon, + 10°
Saw tracks of foxes and of two ptarmigans yesterday. Weather cloudy and snowy. We make for the furthest island in sight, intending to pass east of it, and are stopped by the fog, after a march of thirteen hours, in which we think we have made eighteen geographical miles to the N.E., over the same sort of ground, covered with a dry snow, the lower beds of which show by their colour and consistency that they date at least from last year. Saw several tracks of reindeer and five ptarmigans, three of which we killed The birds are beginning to assume their spring plumage—that is to say, some grey feathers are mingled with the rest. The three we killed furnished a meal for us six, and came very *à propos*, for we are beginning to run short of provisions. It has been decided to-day that we shall make tea only once a-day, in order to economise our spirits of wine, which is also diminishing rapidly. Hitherto our fare has consisted, morning and evening, of about half-a-pound of pemmican, a little biscuit, and a pint of infusion, which we deck with the name of tea; but as that

requires boiling water, we will content ourselves in the morning with a pint of cold water. At the time of our departure from Fury Beach, sybarites that we were, we hardly relished the succulent dish which the Canadians of the north-west call *rababon*—a mixture of biscuit and flour, boiled with pemmican; the flavour of the *réchaud* composed of the same materials, with merely a handful of snow to hinder the pemmican from burning, hardly pleased our palled palates. Ill-natured persons would perhaps say that the grapes are sour, but we have a great contempt for epicures who cannot dispense with such sumptuosities.

30th April and 1st May.—Thermometer, eight P.M., − 22°; midnight, + 16°; noon, + 12°. Thick fog; cloudy weather. At our departure, the north point of the large island is exactly to the east, and a small island quite to the left presents its western part due north of us. The chain of islets before us appears to meet the land more to the west, though we could not make sure of this on account of the fog. The land is very uneven, and we turn to the north-east, in order to pass outside the islands, where we shall keep upon the floe. At five in the morning, between the large island and the islet to the north, our route lies for ten minutes over the ice of a piece of salt water, the opening of which must be to the north. The fog becomes thicker and thicker. Our dogs appear to be affected like ourselves with snow-blindness, for several of them hunt for the route, not seeing it well enough; and after thirteen hours we halt, having marched, as we suppose, twenty miles, but having really advanced not more than from twelve to fifteen.

During the fog, a poor reindeer comes towards us; we lie down on our faces, and as we are to leeward of him, he no doubt takes us for one of the little elevations covered with lichens. The fog is so thick that our sportsmen miss him at a distance of less than thirty yards. He seems astonished at the explosions, and continues to advance towards us, until one of our dogs, breaking from his tackling, darts after him, and both are soon out of sight. Besides missing so fine an opportunity—such, says Captain

Kennedy, as does not happen twice in the lifetime of a sportsman—we are in despair at having lost our best dog; but, fortunately, he comes back panting, after half an hour's run, and, as some assert, with hairs of the reindeer in his mouth. It was an interesting sight to me to see that poor animal rushing apparently upon certain death, raising and lowering his antlered head like a ship in a rolling sea. We had so often seen in anticipation a reindeer killed by us, and cut it up and cooked it beforehand, that the disappointment was sore, and we had not done deploring it all the rest of the day. The only relief to our snow-blindness was to halt a little sooner, both for that purpose, and to take a little rest, of which we had so much need after fifteen days of forced marches.

1st and 2nd May.—passed the Sunday in our encampment; a strong gale blowing from south-east; drift and thick weather. Many of us feel pains in the legs, which we attribute to fatigue, and which I experienced for the first time the day we slept in the open air, not having changed my wet boots and stockings. Mr. Webb has also bluish spots on his legs, which many of our men pronounce to be scurvy. I combat this idea as much as possible, but when we are alone Captain Kennedy tells me he really believes it is scurvy. Alas! as we can do nothing for it, we had better not think of it. For my part, I will not even look at my legs until we are at our journey's end.

I have been unable to take observations with the horizon of ice. All my attempts to form an artificial horizon have failed; whether with the doctor's liniment, with diluted ink, or gunpowder dissolved in water. The breeze and the drift have been insurmountable obstacles. I even tried to freeze water at the place where I wished to observe, in order that the ice might be horizontal, but it was too soft and dull to serve as a reflector. Hence, on such expeditions, I shall always recommend the quicksilver horizon being taken, for quicksilver congeals only at $-40°$ Fahrenheit; and it is not likely that observations will have to be taken at a lower temperature. It is impossible to conceive, without having experienced them, the difficulties encountered when one tries to level the ice-horizon with the thermometer at 20 or 30 degrees; which, moreover, never remains perfectly horizontal, for the snow surrounding the support yields under the weight, or the change of temperature. I used a kind of support, the vertical of which was altering every moment; or, if I employed a massive body, the conduction of heat occasioned a continual movement.

2nd and 3rd May.—Thermometer, eight P.M., − 25°; midnight, − 10°; noon, + 6°. Light breeze from north-west; clear weather. On quitting our encampment, we perceive, to our great surprise, that the supposed islands around us were so many insular masses of limestone, connected together by low spots of land, over which we have passed. Once more, then, we see how deceptive are the appearances of this snow-clad region, especially in foggy weather. We ought probably to have passed outside all these masses to remain on the sea, which is to our right; but to get away from them the sooner we marched due east, and after about five miles, we came again upon the coast to our right. It runs S.S.E., and consists of cliffs set pretty close together. On our left is a very small creek, terminated by a high cape, beyond which we cannot yet see anything; but the portion visible to us, as far as the cape runs, is due north. For the first time since Brentford Bay, we see ice-blocks on the margin of the sea, piled one over the other by the action of the tides. (It must be remembered that Captain Kennedy saw the islets connected together on the night of the 28th; but I was not mistaken in believing that it could not be the coast of Somerset.) We cannot, then, be far from the opening of this inlet, or gulf—a fact we should otherwise be ignorant of, not having had any observations since the 26th of April.

A little above the horizon to the northeast and E.N.E., we see land, which belongs to North Somerset, and which we suppose to be Capes Granite and Pressure. Captain Kennedy remains on the beach to examine the coast, whilst we skirt along it. Numerous remains of Esquimaux encampments of old date, are scattered over this portion of the coast. Sir Edward Parry found similar remains on Parry Islands. There can be no doubt, then, that a portion of the primitive inhabitants have been successively driven out, or forced to emigrate southwards, whatever be the chances in favour of the existence of a native population in the unknown regions adjacent to the North Pole. We ourselves found traces of it all along the coast of Somerset. What can have been the causes of this emigration? The animals that haunt the more northern regions have not deserted them; for the musk ox, the reindeer, and several species of birds have recently been found there; besides which, their annual migrations always take place in that direction. Can it have been caused by the increased cold of those regions? It is hardly supposable that this can have taken place to such a degree as to be noticed by the Esquimaux. Be it as it may, here have passed those wandering tribes whose daily existence is maintained by such fearful struggles; and this gives a more imposing solemnity to the aspect of these stern cliffs, blackened by ages. They seem to look down disdainfully on the miserable ruins which contrast so strongly with their immutable antiquity. The Polar bear and the Arctic fox alone periodically revisit these regions, formerly inhabited, but now completely abandoned, even by the enduring race of the Esquimaux. Here generations perhaps have lived, whom the vaunted light of civilisation never reached; and yet they had, like ourselves, their share of this earth's joys and sorrows.

But, however interesting for us are these relics, and the reflections they suggest, with greater interest and care should we examine more recent traces of habitations; those, namely, of the persons we are come in search of. We are not far from Cape Walker, the first point to be visited by Sir John Franklin, and whence he was to advance south-east Perhaps his ships or his boats have touched not far from us, and we are in a high state of excitement at thought of the discoveries which any moment may now bring us.

Since we reached the coast, we have marched seven or eight miles northward. After passing this cape, the land inclines to N.N.W., as far as another high cape, in the direction of which we march eight miles more. I have had no difficulty in coming to the conclusion that the lands seen by Sir James Ross in 1849, and which he marked as islands, are the same we are now upon; for those of Somerset appear to us just in the same way; the bays and indentations of the coast forming vacancies, which give that continuous line the appearance of a broken chain of lofty capes. The certainty we have now acquired puts us all in good spirits, and will contribute, I hope, to make us find the end of our excursion more agreeable, whatever yet may be its incidents.

I must especially render to two men of our crew, A. Irvine and R. Webb, the ex-sapper, the strict justice of saying that they are men as obedient and as full of alacrity as I ever desire to have under my orders. For my own part, I have nothing but praise to bestow on every one of our companions for their attentions and kindness during the compulsory intimacy of a life in which the petty privations of every day tended so much to provoke selfishness; and never shall I hereafter call to mind the incidents of our voyage without gratitude towards those hearty good men.

3rd and 4th May.—Thermometer, eight P.M., $-12°$; midnight, $-13°$; noon, $+15°$. Light breeze from north-east; weather exceedingly clear. Yesterday the sun did not set, and we are beginning again to have continual daylight, for twenty-four hours. We are now in the humour to devote our earnest admiration to that great orb, magnified by refraction, slowly describing a slight curve above the horizon, as if uncertain whether to sink below it or not; but it has so little time to rest there, that it is not worth its while.

Six or seven miles beyond our encampment, we reach the cape which was in sight to N.N.W., and after ascending it we descry before us Cape Walker, from which we are separated by a wide bay, and Limestone Bay to the east. Possession is taken of this new land in the name of H.M. Queen Victoria, and it is baptised by the name of H.R.H. Prince Albert. The inlet, which extends southward, will bear the name of Mr. Grinnell, the American merchant who sent two ships in search of Sir John Franklin; and Captain Kennedy gives the cape on which we are standing the name of Cape

Bellot. It was his wish, he tells us, to join together on the same day, and upon the first lands he discovered, the names of three members of the three great nations which have taken part in our expedition. The names of the Queen and Prince Albert are hailed with three cheers, and with hearty wishes for the prosperity and union of the three peoples. I would have declined the honour of seeing my name thus joined to that of Prince Albert and Mr. Grinnell, on account of my very humble share in the expedition; but the words with which Captain Kennedy accompanied his declarations left me no choice but to hold out my hand to him with lively gratitude. This day is the more remarkable for me as it is the anniversary of that on which I received the letters of Lady Franklin and Captain Kennedy, in reply to my offer to join the expedition. It reminds me also of another solemnity; and when, in our encampment, my thoughts turn to Rochefort and France, I think again of the events which are taking place there at this moment. What are they? What is their issue? O, that I were a little bird!

Cape Bellot then, since a Cape Bellot there is, is a pretty lofty promontory, formed of a yellowish limestone (of which I carry off a fragment with me*), rugged and bristling like all these cliffs, to which my ignorance of geology does not allow me to assign an exact character. It forms, with Cape Walker, the entrance to a bay, to which Captain Kennedy gives the name of Mr. M'Lean, formerly an officer and shareholder in the Hudson Bay Company, who planned our expedition, and introduced our commander to Lady Franklin.†

In examining the environs we remarked a little cairn, and as it was exactly in a line north and south with the cape, we immediately conceived the hope that it was raised by Europeans; but it was only an old *cache* of the Esquimaux, one of those places in which they bury their seal or whale oil in bags of skin, and cover them with stones.

* It is preserved as a precious relic by Bellot's family.

† "Though, at first sight, it appears that what is here stated is quite irreconcilable with what Bellot has previously related (28*th* and 29*th April*, p. 211), we do not think ourselves justified in suppressing anything. Certain it is, however, that Bellot's account differs from that given by Captain Kennedy."
—*Note by M. de la Roquette.*

From the spot where we are Cape Walker appears very lofty, and has a magnificent reddish brown tint, which makes us suppose it to consist of red sandstone. Another cape, remarkable for the same colour and for its height, at the bottom of M'Lean Bay, is named Cape Barrow, in honour of the son of Sir John, to whom our expedition owes so much, and who was lavish of polite attentions to me in England. Captain Kennedy tells me that Lady Franklin instructed him (which I knew) to give the names of several of my friends to part of the coasts we should discover. I declined this with thanks, for I deem it more becoming to think first of those who have contributed to the fitting out of the vessel; and that, even for that purpose, our discoveries are unfortunately too few; otherwise, the names of M.M. Bonnaudet, de Lescure, and Desfosses, to whom I owe my position and what I am, would have been the first to present themselves to my memory. The friends of my heart, too, should certainly have had a place on the line of our discoveries, and my mother, and my sister Estelle, and others. By-and-by, perhaps; who knows? I here insert the passage in Lady Franklin's written instructions relating to names:—" M. J. Bellot and his friends; let all the French names be upon one portion of the coast, in order the better to attract attention."

A little low island, or rather a reef, covered with ice, stretches towards the middle of the entrance; and just as we are about to pass it a great bear escaped from it, but out of gun-shot. We also see at the bottom of the bay an inlet, the extremity of which we, no doubt, crossed on Saturday before encamping. In spite of our efforts, the heat had become too oppressive before we could reach that Cape Walker, to which is attached a melancholy celebrity, and we had to halt some miles south-east of it on the ice of the offing. Postponing our researches, then, till to-morrow, we hastily throw up some pieces of snow between us and the wind by way of encampment.

4th and 5th May.—In the evening, Messrs. Kennedy, J. Smith, and I, proceed towards the west. Mr. Smith bears a little to the east; and, whilst Captain Kennedy is examining the beach, I keep a little wide of it on the ice, to see if there is not some mast, or other signal, fixed on the top of the cliffs. After having passed the reddish faces of Cape Walker, the land sinks gradually, and inclines insensibly to the west. The beach is very narrow, covered with coarse sand, pebbles, and fragments of limestone. After three hours and a half of very rapid walking, we meet again on the coast. Mr. Kennedy, who was the furthest in advance, had climbed a hill, whence the land appeared to him to run in a south-easterly direction. To our immense astonishment, not the least trace either of Sir John Franklin's vessels, or even of those of Commander Austin. The Europeans appear not to have touched yet at this part of the coast, for not one of them would have done so, especially the last ships, without erecting some signal, or leaving some vestiges of their presence.

We calculated that we were then six miles west of Cape Walker. The weather was very unsettled when we quitted the encampment, and the mist had become so thick, and the wind was getting up so strong from the south-east, that, whether we would or not, we were forced to think of returning. Though we had taken by compass the bearings of the spot where our baggage lay, we lost our way in the fog, our snow-shoes having left no track on the unsnowed ground. Very fortunately, the men left in the encampment had thought of firing shots from time to time, and as the wind blew towards us, we heard the reports, and were thus relieved from an embarrassment which might have become serious. I know not how things will be regarded, but I am afraid Captain Kennedy will be blamed for not having left anything to mark his visit, to give information to those who might come to the spot, if the ships of the Arctic squadron are still in the vicinity. I proposed to him to go with Mr. J. Smith, as soon as the weather cleared, to the foot of Cape Walker, erect a cairn, and deposit in it the documents he might think necessary; but, as he has lost a small note-book and some papers, which were to be deposited in the cairn, he thinks we may leave the thing undone.

We have but sixty pounds of pemmican

left, without biscuits or tea, for ourselves and our dogs, and we are still a hundred and twenty miles, in a straight line, equivalent to a hundred and fifty in reality, from Port Leopold, the only spot where we can find provisions. We resolve, with grief, to resume our course eastward—I say in all truth with grief, especially in the present state of things, no ship appearing to have reached this point. Only a few days more, and we should have been able to push on fifty miles further west; and who can say what we should find there? But Captain Kennedy could not come to any other determination without plunging into fresh danger. The hundred and fifty pounds of pemmican and biscuit we had when we left Brentford Bay, supposing a pound of biscuit to be equivalent to one of pemmican, were to last us only twenty-two days, at the ordinary ration of two pounds for each man and each dog. We have lived thirty days upon these provisions; and, if we should find the ice broken at the entrance of Grinnel Passage, and were obliged to turn south again before crossing that passage! O, if we had crossed Somerset Land from Creswell Bay!—we should have been here perhaps a fortnight sooner, and then—— But who could have foreseen what has happened to us, and how we were not to find things as they had been described; how we were to be retarded by fogs, &c.?

We are forced by the bad weather to build our snow-hut, or rather our usual half-hut; for within the last fortnight we have found it more expeditious to erect merely a circular wall, which we cover in with a large piece of canvas. On our return along the coast we found foot-prints, on the origin of which we were beginning to speculate, when at last we perceived by certain marks that they were our own foot-prints, made a few moments before, but now covered by the drift. This shows us once more how reserved one ought to be in making assertions as to tracks of this kind. We also saw tracks of reindeer of last autumn; and Mr. Smith thinks he recognised, eastward of Cape Walker, remains of an Esquimau encampment, but of very old date.

5th and 6th May.—Thermometer, six P.M., + 17°; midnight, + 4°; six A.M., − 4°.

The temperature is rapidly rising under this thick mist, and we are wakened by the snow melting through our canvas roof and dropping on our blankets. Fortunately, a strong north-west wind partly disperses the fog, and at six P.M. we set out for Cape Pressure, on the opposite side of the inlet. After fifteen hours of hard work, hauling the sledges over the ice-blocks and soft snow which formed our road, we encamp again in sight of Cape Walker, and the lands to the south-east. Those to which we are going, appear from time to time in the openings through the fog. Our dogs are become so voracious that they eat all the leather they can find— our gloves, boots, snow-shoes, &c., and we are obliged now to take all these things with us into the interior of our encampment. If Grinnell's Inlet is not always entirely open in summer, it is certain that the entrance at least is so from time to time, as appears from the rough ice in it.

6th and 7th May.—Thermometer, six P.M., $-2°$; midnight, $-10°$; six A.M., $-4°$. Clear weather; light breeze from the west. About midnight we lose sight of Cape

Walker, which disappears in the fog due west behind us; but we perfectly distinguish Limestone Island to the east, and Cape Pressure. We start a bear, larger than any we have yet seen. He rubs his eyes, and after a long circuit, places himself athwart our route, as if intending to dispute our passage. The animal has probably never seen man before, and takes us for some sort of biped on which he intends to gorge. Seeing him ready to give battle, we prepare all our arms, and four of us march straight up to him. He lets us get within thirty yards, but our guns giving him perhaps instinctively to divine what are our intentions, he makes off with such speed, that after a quarter of an hour we are obliged to give up the chase. His imprudence deserved a more severe chastisement, and if we had killed him we should have returned to Cape Walker.

At seven A.M., we witnessed the most splendid meteorological phenomenon. The sun, about 24° above the horizon, is surrounded by a perfectly formed halo, with two horizontal, and one lower vertical, parhelia. A horizontal crown of white light, or parhelic circle, passes through the centre of the sun, and stretches athwart the sky parallel to the horizon. On this crown two brilliant discs of white light, or parenthelia, form with the sun an equilateral triangle. Above the halo, a portion of circumzenithal arc, of about 40°, exhibits all the colours of the rainbow. Some portions of the lower halo were also visible. Unfortunately, Captain Kennedy would not stop to take the dimensions of this phenomenon, and it was too late when we reached the encampment. The temperature was − 3°, and the air was filled with snowy particles, quite visible to the naked eye.

In the morning the sun was about 6° above the horizon; a part of the halo appeared with the parhelia, which projected long bands of white light, divided in two by a solar cross, which widened towards the base of the vertical column. Other matters more immediately claimed our attention; but, as the reader sees, almost every day presented to us an interesting meteorological spectacle. It seems, indeed, as though Nature reserves her most magnificent phenomena to compensate the toils of the traveller in those icy regions. O God, how beautiful are thy works!

Our route to-day was a little better than yesterday; nevertheless, Captain Kennedy and I have been again obliged to go a-head, beating the route with our snow-shoes, in order to compress the snow, or else the sledges would sink in it. We also found tracks of bears so numerous, and in all directions, that this inlet quite appears to us to be a rendezvous for those animals. We encamp on the ice two miles west of Limestone Island.

7th and 8th May.—Thermometer, six P.M., $-1°$; midnight, $-8°$; six A.M., $-3°$. Wind south-west. After passing the island, in order to shorten our route, we try to encamp in the direct line to Cape Rennel; but, after stumbling about among the ice-blocks and soft snow of the offing, we are obliged to return to the coast. Though the thermometer is not very low, and we should now be accustomed to cold, we feel it keenly, perhaps in consequence of the diminution of our food; for now we must content ourselves with a little bit of pemmican—barely five or six ounces—which we dissolve in boiling water; and this thin broth, without bread, is not very nourishing for men who have to toil so hard. We pamper ourselves all the more in imagination, and we cannot help laughing as we remark how our conversations these last few days run upon the dishes that each likes best, and on what we will do at Port Leopold, where provisions are abundant. It is not, however, on sumptuous so much as on plentiful food—on our fill of pemmican—that our thoughts run; and we rummage our pockets for any crumbs of biscuits that may have been forgotten in them in times of abundance.

8th and 9th May.—Thermometer, six P.M., + 3°; midnight, − 1°; six A.M., − 2°. A halo and two parhelia. Cloudy weather; snowdrift. Light breeze from south-west, which makes the cold seem excessive. After fourteen hours of rapid marching over the beach and the ice, we encamp two miles east of Cape Rennel, having crossed Cunninghame Inlet at four A.M. Saw several tracks of bears. It is beginning to be time that we were arrived.

9th and 10*th May.*—Thermometer, six P.M., −2°; midnight, −9°; six A.M., +1°. Clear weather; light west wind. We make straight for Cape Maclintock, and encamp west of Gurney's Bay, after a march of thirteen hours over the beach and the ice. We still cross numerous tracks of bears, coming and going in all directions. Four of us feel pains, the cause of which no longer admits of doubt, though we avoid speaking on the subject. I have discovered on my legs the little blackish spots, which are the surest signs of scurvy. But what can we do?

10*th and* 11*th May.*—Thermometer, six P.M., −1°; midnight, −4°; nine A.M., −1°. Clear weather; light west wind. We continue our course towards Cape Maclintock, and encamp three or four miles west of it, after twelve hours' march. From the beginning of the evening we have had Leopold Island and Cape Clarence in sight. Nothing can be more magnificent than the jagged strata of limestone composing the island, and reflecting the rays of the western sun; but who among us thinks of admiring these beauties? Cape Clarence and Port Leopold have now but one meaning for us: they are a vast hotel signboard, with the inscription, "*Table d'hôte and board!*" Above the horizon, we see to the north the other shore of Barrow's Strait. It is calm; the thermometer is at 18° below zero centigrade. In order to be ready the sooner to set out for Port Leopold, we spread a tarpaulin on the ice and our blankets over it; by that means we gain time, and dry our baggage, which will be so much the lighter. In order to have something for breakfast to-morrow, we give the dogs our old mocassins, torn gloves, and a piece of bison's skin with the hair on, and keep their ration of pemmican for ourselves. This is the first time they have not had the best share. Our provisions being thus exhausted, we must make a long stage to-morrow; and it is arranged that Messrs. Kennedy and Smith shall go on before the rest to Port Leopold, prepare our quarters, and meet us with some food. The other three men and I are to follow as fast as we can with the sledges and the baggage. Saw numerous tracks of bears.

11*th and* 12*th May.*—Thermometer, six

P.M., − 4°; midnight, − 9°. Stiff breeze from the west; cold moist weather. At three P.M. we start. Messrs. Kennedy and Smith are soon out of sight. After we had gone about three miles, we saw them, to our surprise, halting at the foot of a cape immediately west of Cape Maclintock, and concluded that they must have something of importance there. In fact, they had discovered a depôt of provisions left by Sir John Ross in 1849, for Sir John Franklin, as appears from a note enclosed in a tin case. There were several barrels covered with snow or ice, for the sea-water had reached them, and containing a considerable quantity of biscuits, sugar, pemmican, chocolate, and preserved meat. These, with wood and stone-coal, which we also found, left us nothing to desire. Our dogs and ourselves took an ample instalment at once, and we set about building a snow-house, in order to rest a whole day. Two of our men were appointed cooks, and under the beneficent influence of some hot chocolate, we soon forgot our late privations. We even forgot the rules of prudence and sobriety, for which several of us were punished in a comical manner. We unanimously bestow on this nameless cape the name of Cape Mercy, from feelings of thankfulness which need no comment. It would be hard, however, to say how much we regret that we did not meet with this depôt at Cape Walker or Limestone Island.

12th and 14th May.—Thick and snowy weather. Light west wind, blowing in gusts. We spend these three days in eating, drinking, and sleeping; drinking, sleeping, and eating; sleeping, eating, and drinking, without regard to consequences. Captain Kennedy was so unwell on the second day as to render another day's halt necessary. Besides, we are no longer in such a hurry to reach Port Leopold.

14th and 15th May.—Thermometer, six P.M., $+5°$; midnight, $+2°$; six A.M., $+1°$. The weather clears up on the afternoon of the 14th, and we set out, feeling much greater fatigue than before our three days' rest. The lands to the north are very distinctly visible, and as we approach Cape Clarence we see unequivocal signs of the presence of running water north and east of it. About three o'clock we get sight of a signal-mast set up on the southern shore of Barrow Strait in 1849; and, crossing the low tongue of land which unites Cape Clarence to the mainland, we are at last on the ices of Port Leopold. In October last we left the sloop, which then served us for a dwelling, completety covered and close, and we expected to find ready-made quarters in it; but, to our great dissappointment, an infernal bear had broken down the roof, and the sloop was filled with snow so compact and hard, that to clear it out would be a more tedious work than to build a snow-hut. We set to, then, at the latter, repeating Hamilcar's imprecations against the whole ursine race.

15th and 16th May.—Strong gale from the north, and snowdrift. Decidedly we are all more or less attacked by scurvy. It is now useless to conceal the fact from ourselves, and all we have to think of is how to get rid of it as soon as possible. We are very uneasy about those on board, who are most probably in the same plight as we are. No doubt they, too, are very anxious about us, our absence having been longer than they calculated. For these reasons, we should be glad to hasten our arrival among them, but there are not two of us now capable of reaching Batty Bay: besides, we have here all the requisites for use in greater abundance and better condition than on board; that is to say, lime-juice, vegetables, fresh meat, pickles, sweetmeats, and other refreshing things. Captain Kennedy is, therefore, resolved that we shall remain here until our health is better. We build ourselves a more spacious house, and set up a tent made of scraps of canvas for a kitchen, in order to keep apart from sleeping-place the causes of humidity, which alone can give us the scurvy. For the same reason, we return to the usual course of sleeping by night, so that we may dry our things by day.

After the first junketing, satiety fortunately subdued that everlasting appetite that threatened to absorb every other faculty, and we now enjoy in moderation the plentiful resources at Whaler Point. After fifty days, during which it was hard to divert a drop of water from the indulgence of our inextinguishable thirst, it is no small delight to be able to attend to personal cleanliness: in a word, everything yields us physical enjoyment. This hinders us, perhaps, from feeling as we ought what we owe to Providence. What would have become of us if the scurvy had thus manifested itself at Cape Walker, or further still? There can be no doubt as to the consequences: and during the last few days, too, when cold and hunger were diminishing our strength, the wind was always west, so that we had it at at our backs; and the ice, which we might have found open between Cape Walker and Limestone Island, afforded us a short and easy passage. I do not think I exaggerate the dangers we have escaped; but where is not God's finger to be seen, even in the slightest things?

17th May.—Thermometer, six A.M., $+17°$; noon, $+12°$; six P.M., $+15°$. Strong breeze from the north, accompanied by the everlasting snowdrift. We have finished our snow-house, and find it so comfortable that we regret not having done the same thing at Fury Beach; for a snow-house would certainly have been more comfortable than one of canvas, in which we were blinded with smoke; certainly, it is warmer. One of the stoves left by Sir John Ross is fixed in our tent kitchen, and serves for all our wants. The bear killed by Captain Kennedy in October was put into one of the steam-boilers of the sloop. One of the doors has been torn down, and the tracks left on the snow show that a bear was the author of this misdeed; the circumstances of which lead us to think that there is as much cunning as strength in that clumsy animal.

18*th May.*—Thermometer, six A.M., + 22°; noon, + 12°; six P.M., + 17°. Light west wind; weather a little clearer. We spend the day in spreading out and beating our things, in order to dry them well. In a short walk to Cape Clarence, we see a channel of running water, which extends eastward as far as we can see, and to the south farther than Cape Seppings, which is about ten miles off. We are all furnished with crutches, and look very like a detachment of invalids; but we try to keep up our spirits and our activity. Movement and exercise are the chief remedies against scurvy, and every means should be employed to rouse those who persist in lying down: the mus cles contract so soon, and so obstinately, that the same obstinacy should be used in combating the malady. Thank God! I am still of good courage; and I hope that a few days of our present regimen and a little exercise will set me up again. My legs are much swollen, especially under the knee, and the black spots have not yet disappeared. Messrs. Kennedy and R. Webb are the greatest sufferers; and I am afraid that in the former scurvy will unite with rheumatism to do him mischief: but he yields to none of us in activity; and his ceaseless efforts to encourage our patients—above all, by his example—will, I hope, have a good effect upon their health. We have got over the worst of this work, I suppose, and I think I have acquitted myself becomingly; not but that inwardly I was afraid, on two several occasions, that my courage would fail; but, fortunately, at the critical moment I recollected my position and my character. Thanks to Heaven for it! for I had a hard apprenticeship to undergo; and all here, except myself, had an experience of the toils of such expeditions, of which I was entirely devoid. What mental torments, too, had I to endure, in addition to the physical difficulties! but I kept these momentary conflicts to my own breast, and no one can say that a French officer gave way when others stood firm.

19*th May.* —. Thermometer, eight A.M., +22°; noon, +27°; six P.M., +28°. Light west wind; clear weather. The snow is melting all round us, and the opening in the ice appears to extend more towards the south-east. We spend the day in getting out of the snow the gutta-percha boat left here in October; for the sight of running water has suggested to us the idea of boating to Batty Bay, which would spare us much fatigue, and enable us to bring away with us a supply of lime-juice, of which our men on board must have need, for the ship's stock is both bad and scanty. To avoid wet from the snow that melts in the sunshine, we revert to our former plan of working by night. On the evening of this day the breeze freshened, and the thick snowdrift forced us back to our quarters, in spite the desire and need we have of bodily occupation.

20*th and* 21*st May.*—Thermometer, six

P.M., + 11°; midnight, + 8°; six A.M., − 15°. Light north wind; cloudy weather. We try all means to find work for our arms and legs, and fall to at the sloop to clear it of the snow with which it is filled, and which is so hard that we are obliged to cut it with axes and saws. This job may be useful, as relieving the sides of the sloop of a pressure of many tons' weight of snow, the roof rising five or six feet above the boat.

21st and 22nd May.—Thermometer, six P.M., + 11°; midnight, + 15°; six A.M., + 25°. Light north wind; snow from time to time. We have finished the excavations begun yesterday; we then set to work to mast and rig our boat, which had left the schooner on the 9th of September, without having its sails bent. We amuse ourselves beforehand with the surprise of our shipmates when they see us arrive in a boat, and from Port Leopold; for, if they expect us at all, it is a thousand to one against their guessing that we shall come from the north. It is rather remarkable that we have seen no animals here, except two crows, which let us approach them very near, and seemed more surprised than frightened at our presence. We recollect that this was the only species of bird we met in our winter excursions. On what do they live at that season of the year? The weather again turns thick and snowy on the morning of Saturday; but at last our preparations are nearly complete, and there is a manifest improvement in the state even of the worst invalids. It is decided, then, that we shall start on Monday morning, if it pleases God, and the weather permits.

22nd and 23rd May.—Thermometer, eight A.M., + 22°; noon, + 25°; six P.M., + 26°. Strong west wind; thick weather; drifting snow. We rested on Sunday, and for the first time, after a long interval, we joined in offering our thanks to God, and imploring new favours from His inexhaustible goodness.

24th May. — Thermometer, eight A.M., + 24°; noon, + 28°; four P.M., + 25°. Strong breeze from the west; snowdrift; thick weather, confining us to our dwelling. Captain Kennedy left, along with the other papers deposited here by the commanders of the several expeditions, a note describing what we have hitherto done, and stating that, after we rejoin our vessel, our first business will be to examine the bottom of Grinnel Inlet; after which we shall probably visit Wellington Inlet. This, I suppose, has reference to a plan which I proposed to Captain Kennedy, but of which it appears he had previously thought. When the question of the excursion to Cape Walker was under discussion, my advice would have been to go up the coast a little way north-east; then down it again south; up it again northward to the height of Cresswell Bay, where we should have found the cairn raised by Sir John Ross in 1849; and to have crossed in the direction of Fury Beach. I then thought that the opening at least of the inlet—that is to say, Cape Walker—and its environs had been examined by the Arctic squadron of last year, and that in that case it was better for us to explore the bottom of the inlet, and reconnoitre the coast as far as Four Rivers. However, in the actual and apparent state of things, if we alone have visited Cape Walker, I doubt not but that this will give much more satisfaction to Lady Franklin, and to the public in England; but I did not insist, upon his observation that this would be done by boats during the summer, which I believe to be hardly possible now, from what we have seen.

25th May.—Thermometer, six A.M., +18°; noon, +27°; six P.M., +28°. Still confined by the same weather; strong breeze from the north; from nine to six, snowdrift and snow. Our poor dogs seem to suffer from the drift and the melted snow, which fastens in their long hair. We remark that they approach us, but we cannot guess the reason until we see them ridding each other of their icy encumbrances.

25th May.—Thermometer, eight A.M., +25°; noon, +24°; six P.M., +24°. The wind having fallen, we put everything in order, and in several trips convey our boat and our baggage over half a mile to the edge of the ice. We all feel a childish joy at being once more on the water, and seeing our boat glide between the floating masses of ice, which the wind and the surge are continually detaching from the land-floe. The scene is, moreover, animated by the cries of sea-birds which we disturb, and which we recognise with shouts of joy: common gulls, the graceful diver, the gluttonous mollyknock, the kittiwake, the eider-duck. Big seals, with bristly moustaches, pop their heads out of the water to look at us, and disappear instantly. One of us even thinks he perceived the eddy caused by a whale. We expected to reach Batty Bay to-morrow morning, but the ice barred our passage fifteen miles south of Port Leopold; so that we had again to unload all our baggage on the ice, and carry it and our boat to the foot of a wide ravine, where we spent the night and the following day. After all, the little way we have made is so much gained. What vexes us is, that we cannot carry with us four barrels of lime-juice, and three of conserves, which we had intended for the schooner, and the want of which we shall regret if the scurvy is on board. On the way we passed close to a bear drifting on an ice-block, and no doubt in pursuit of seals.

28th and 29th May.—Thermometer, six P.M., + 24°; midnight, + 24°; six A.M., + 32°. In the state we are in, it is essential that we should reach the ship as soon as possible; so, leaving behind us the boat and all but the indispensable portions of our baggage, we set out on Friday evening on foot. Lightly as the sledges are loaded, they make but slow progress through the slush that lies above the ice. Near Elwin a bear appeared for an instant, but ran off with such speed that we could hardly believe it was a bear until we saw its track. It is probable these animals often fast during this season, for the seals are exceedingly shy, and never let us come near them. The floe measured at one of their holes was but two feet and a half thick. At five in the morning we encamped on the ice of the offing, five miles south of Elwin Bay. The thermometer is at 0° centigrade, and we dispense with any other shelter than our blankets and the sky.

29th and 30th May.—Thermometer, six P.M., $+29°$; midnight, $+25°$; six A.M., $+26°$. Fine weather; light breeze from south-east. We set off again in the night, and it is fortunate we do so; for the beach, over which we are obliged to march on account of the bad state of the ice, is strewn all over with huge stones, detached from the cliffs by the snow-water and the heat of the days, some of them weighing many tons. In two places many marks of avalanches show how unsafe it would be to pass by day along the foot of the cliffs. We see three bears; one of them comes so near us as to be hit with two balls, which knock him down; but, before we can despatch him, he makes off, leaving a long track of blood, which makes it probable that his wounds are mortal.

At the entrance of the bay, we remark with pleasure wide clefts in the ice; signs of certain dissolution; and, as we hope, of speedy deliverance for ourselves. At five in the morning we are on board; but a foreboding of some bad news hangs over us and checks our glee. Captain Leask is the first whose hand we grasp; and amidst his congratulations, before he has said a word of our shipmates, we have guessed that our fears were well-grounded, and all are not as well as we should wish. But who would dare to murmur, after so many bounties of Providence? Here we are, once more united! What God does, is well done!

CHAPTER IX.

RETURN ON BOARD.

31st May.—As I said before I wrote the journal of our excursion, the majority of our men have been attacked by scurvy; only the captain, Mr. R. Anderson, the third officer, and Messrs. Grate, G. Smith, and Linklater, are exempt from it; many of the rest are in a frightful state; the doctor and Mr. Hepburn are very bad. Their small stock of lime-juice was quickly consumed, for they found the barrel congealed, and the essence of the juice concentrated into a ball in the middle, the rest being as insipid as common water.

1st June—It will never do to consider ourselves very ill alongside of those who are much worse than ourselves; and I proposed to Captain Kennedy to lead a party of our able-bodied and convalescents to the place where we left the lime-juice. We set out on Tuesday evening, under a north wind, to which we could not have exposed ourselves with impunity a few months sooner, but for which we don't care much now. An Esquimau sledge, and our five dogs, rapidly convey our light baggage; and, after having slept by day on Wednesday on the ice, we arrived on Thursday morning at our depôt, where we were detained the next day by a gale. We were to try to take with us the jollyboat, or *youyou*, which was left on the coast in October, 1851; but one of its sides had been torn by a bear, and bears marks of his teeth, which inspire us with thoughts of vengeance, if we have opportunity. Having dragged the boat to the foot of the ravine, where the gutta-percha boat had been left, we are obliged to leave it there, on account of the bad state of the way, which we much regret, for it is the most useful of our boats at a landing. We see a flock of blue weavies, or Canada geese, flying north, being, no doubt, the vanguard of a more numerous army. Our men began to scream out with all the force of their lungs. I could not tell what it was for, until, to my surprise, I perceived that the whole flock repeated the same cries, wheeling round and round over our heads, and at last resumed their flight without being molested, our gun not being in order. Anderson says that these birds winter in numerous flocks in the savannahs of Florida and Arkansas, and begin their migration northward at the first melting of the snow, between the 20th of March and the end of April. I am told that nothing is easier than to lure them with rags made up into the shape of birds.

On the evening of Friday, the 4th, we resume our route. It is easy to see that the ice is rapidly thawing, and is already in a very different state from that it presented not a week ago. We saw four bears on our way, and one came sniffing our baggage twenty paces from our tent, while we were asleep under it. Our weary dogs did not awake, but two young dogs that followed us from the vessel at last gave the alarm, and put the bear to flight. We often laughed, at the beginning of the cruise, at the fear with which bears inspired other voyagers; but more complete experience has shown us that at least it is not prudent to neglect all caution with respect to them, particularly at the time when the females have young ones, and prowl in search of food for them. One of those we saw was lying on the ice, watching for the appearance of a seal. It is said that, keeping back their bodies, they encompass the hole with their paws, and catch the animal when it rises above the ice. Mr. John Smith tells me he saw a bear jump into the water, and come out with a seal in its mouth, which would show that they are swifter swimmers; but I am not sure how far the fact is authentic. Seeing many sea-birds hovering and screaming over a carcase, we imagined it was that of the bear of the 30th, which had died of its wound; but it was only the skeleton of a seal, which a bear had dragged more than a mile, as appeared from a bloody track, in order to devour it more at his ease. The marks on the snow showed that the seal must have fought hard for life.

6th June.—At last we arrived on board this morning, much fatigued, but no longer ill from our exertions, and delighted at having done something useful. A herd of sixteen seals, lazily basking on the edge of a cleft in the floe, dived in at the sound of our sledge. We saw only one fox-track—those animals being now no longer common—and one white owl.

I will now insert an account of what passed on board during our absence—that is to say, since the 17th of March, when I left the vessel, with a stock of provisions, to go to Fury Beach.

CHAPTER X.

ON BOARD DURING OUR EXCURSION.

18th March to 16th April.—The same series of gales and snow-falls which we encountered took place also at Batty Bay. The heat of the sun occasioned notable changes in the state of the ice and the cliffs, which, the doctor says, emitted dull, rending sounds. Frequently the deck was covered with water streaming from the woollen awning. The ship's company set traps for foxes, hoping, in the absence of most of the dogs, to catch some of them alive, to be let loose afterwards with collars and inscriptions round their necks; but the two young dogs left on board were always at the traps before the men, and when they were shut up in any of our snow-built stores, they soon worked their way out; so that only two foxes were caught, and those dead. Hunting and shooting, the only recreations of our shipmates left on board, were unproductive, with the exception of two ptarmigans killed on the 18th of March, and four foxes killed at the end of that month, besides the two that were trapped. One of them appeared better fed than the rest, but its stomach contained only scraps of canvas and of old gloves. One or two crows hovered round the ship, and, as usual, were too wary to be killed, in spite of the desire we have always felt to know on what they feed. For the first time, on the 10th of April, saw tracks of a hare and two snow buntings. On the 14th, Mr. Hepburn began to show symptoms of scurvy, which would not have been formidable, but that the only barrel of lime-juice was frozen, and so insipid that a whole bottle was not equivalent to the usual dose of half an ounce. On the 16th, Mr. Anderson returned on board, with the men we left on the 6th at the entrance of Brentford Bay. Two only were slightly indisposed; the rest had only suffered since the 6th from inflammation of the eyes.

16th to 27th April.—The doctor had been ordered by Captain Kennedy to carry some provisions to Somerset House against our return, and then to go and examine the bottom of Creswell Bay, to see if there was an opening to the west; but he was detained several days by the illness of Miller, who, as well as the carpenter, was afterwards attacked with scurvy. Mr. Hepburn was getting better, and the doctor resolved to start on the 27th, chiefly that he might bring back some lime-juice from Fury Beach, which was the more wanted because a species of sorrel, called by the whalers scurvy-grass, which had been found in plenty by our predecessors, had nowhere been met with in the neighbourhood of the bay. In the interval, the rest of the crew were employed in clearing out the hold and knocking down the ramparts of snow round the ship, as they seemed to keep up the dampness on board; and the ice was then scraped out that had formed on the blocks, the bulkheads, &c. I think I mentioned that in our various snow-houses built on the ice, water had been seen filtrating through the snow by the end of February. As soon as the sides of the schooner were laid bare, the same melting ice was seen on one of them, though its aspect was north, on the 20th of April.

27th April to 21st May.—After the doctor's departure with four men, nothing worth noticing appears to have occurred till his return, the principal occupation being re-stowage, cleaning and drying the vessel. A bear was killed by Captain Leask quite close to the schooner, but no other animals. On the 21st three of the doctor's party returned, saying that they had left him at Fury Beach with the fourth man, who was too ill to travel. They themselves were severely attacked by scurvy, which, I forgot to say, continued its progress on board. Mr. Anderson, the third officer, Grate, the boatswain, and Glennie, immediately started to bring back the doctor, which could not be done by the first three men, in the weak state in which he and they were. I borrow from the doctor's journal the account of his journey to Cresswell Bay and back.

They arrived at Somerset House on the morning of the 3rd, after great difficulties caused by the bad state of the snow, in which they sank up to the knees, and often to the middle. One of the sledges, which had been damaged by the hummocky ice, and the tent, which had been gnawed by foxes, had to be repaired. They were detained by bad weather until the 4th of May, when they encamped near our first snow-house, having found the ice so smooth and clear of snow that they could hardly keep on their feet. Wednesday, the 5th, followed the outline of the land on the floe, where the snow was a foot thick. This day they suddenly saw coming towards them seven bears, whose presence they could well have dispensed with, as they had but one gun. They kindled a fire in the hope of scaring them, reserving their musket as a last resource; but fortunately four of the bears ran off, after coming within 200 yards, snuffing the scent of their track, whilst the other three kept at a distance of a quarter of a mile. "On the 7th of May," says the doctor, "we were abreast of the high land which forms the extremity of a chain of hills running north and south. From the foot of these hills a very low plain of considerable extent stretches to the shore, which first runs south, then towards

———* at the head of the bay, in a western direction. This evening saw the land lying all round the bay, and, consequently, I do not hesitate to say that there is no passage to the west of the bay." The doctor says that his observations were posted down rather hastily, and that he must revise them at more leisure. At midnight, on the 10th, they arrived at the spot where we halted on the 5th. Saw two bears.

Tuesday, 11th, encamped at the same place as on the 4th; boiled water for tea partly with wood, and partly with moss. 12th, reached Somerset House, and found among the things left there by Captain Kennedy and our party, a barrel opened by the bears. M'Currus complained all day of his legs, and the doctor found his ankles swollen. The weather forbade their return on board. 13th, a bear. 14th, Matheson also showed symptoms of scurvy. 15th, the doctor is decidedly affected in that way. On the 17th, Linklater is attacked, and Gideon Smith alone remains exempt from the malady. After a whole week's gale from

* Blank in the MS.

N.N.W., they start on the 19th, taking with them a small quantity of lime-juice; but are obliged to return to Somerset House, finding it impossible to proceed with their baggage. They resolve, therefore, to leave it behind, and take only their blankets and as much pemmican as each can carry in his pockets. On the 20th they make a fresh attempt, but the doctor is obliged to retrace his steps, and to send back to the vessel those who have strength enough to reach it, which they did on the 21st, having marched without a halt. Two bears were seen squatting themselves down by a barrel of sugar, and it took no less than four pistol shots to scare them away. Next day another bear was seen rummaging among the casks at the door of the house, through a chink in which the doctor shot him dead.

The doctor was conveyed on the sledge, not without difficulty, on account both of the stones falling from the cliffs beneath which they were obliged to march, and of the soft snow; and he and his companions arrived on board on the evening of the 27th. There were now seven patients ill of the scurvy, and the boatswain was slightly so: the four others, including the captain, were well; but the vessel was without the essential remedy, lime-juice; besides which, they were beginning to be very anxious on our account, and depression of spirits is especially pernicious to the scorbutic. Their anxiety, however, having been relieved, they soon rallied; and, with the help of a little lime-juice brought by us, they have manifestly improved within this week.

8th June.—We are all now put under a course of lime-juice, our stock being sufficient. The convalescents take two ounces a day, and the invalids from four to eight— the latter quantity being the maximum which the doctor thinks it imprudent to exceed; for the acid has a very debilitating effect, especially on the stomach, as we found at Port Leopold, where we used it a little too freely. The doctor has also prescribed for us a special diet, consisting of vegetables, fresh meat, rice, potatoes, and dried fruit in puddings. The mustard and cress which we grow near the kitchens, are also an excellent anti-scorbutic; Sir Edward Parry used them in the winter of 1819—20. The effect of regimen is helped by extreme cleanliness, a good dry bed, and as much exercise as possible. Our scorbutic patients are sent abroad, and those who are least affected do little jobs on board. It is easy to mark the progress of convalescence in those among them who have most energy of mind, and impose upon themselves a certain number of hours for walking and exercising daily.

Our vicinity is from time to time enlivened by the visits of the snow-bird, which comes as familiarly as our sparrow, flutters over the deck, or hops about near the ship, picking up scraps and crumbs, and only flying off when our young dogs pursue it, and then returning immediately. I cannot help connecting this desire for communication with our species, displayed by some animals, with the pretty fictions of the metempsychosis.

On the 4th of this month the tilt-cloth was at last taken off the vessel, and the air, as well as the light, now circulates freely. This might certainly have been done sooner;

and, generally speaking, I think that retaining the tilt-cloth longer than is strictly necessary is voluntarily prolonging the duration of winter; but it served to shelter our invalids on deck, their only place for walking. The powder, which remained during the winter in our snow magazine, has been reshipped in perfectly good condition. The collection of stones for ballast, placed near the ship in February, has sunk in the ice, and we are beginning the job again; and it is no slight one, since we require not less than thirty tons. The vessel is still imprisoned; it does not float now, but is supported by the ice around it. The joyous spring has but begun for us, but with health and the glad sunshine our vigour is redoubled, and we are busy with preparations for our departure from the bay. We fill our water-casks from large pools of fresh water which have formed in the snow over the ice, and this spares coal and labour. All the rubbish thrown out of the ship, whatever be its nature—wood, metal, &c.—sinks in the snow, and, a little more slowly, in the ice itself: even the seals seem voluntarily to serve us in that way, by the many holes they make in the ice for the purpose of coming out to bask in the sun. The thaw then is going on rapidly, and it is very probable that ere long it will be impossible to reach land dry-foot. The land encompassing the bay is cleared of the snow that so lately covered it, and the flat grounds at the foot of the hills have been converted by the melting of the snow into impassable quagmires. Flocks of geese have passed us, but out of shot. Captain Kennedy killed a seal to-day, which, with Mr. Leask's bear, fills our larder with game. Our men, encouraged by example, are beginning to forget their prejudices; but it is rather late for that.

9th June.—To my great surprise, Captain Kennedy this day communicates to me an idea which must have sprung up in his mind within these last few days—namely, that we crossed Brentford Bay the very day we thought we encamped at its entrance, and on the north coast; that is say, that we encamped at the bottom of the bay, crossed next day the narrow isthmus which divides Somerset and Boothia Felix, and that the sea traversed by us on the 8th and 9th of April is the sea of Sir James Ross, to the west. It is impossible for me at first sight to find objections to his idea, on account of our negligence in determining the route we pursued; we ought, in fact, to have taken the bearings by compass every time we changed our direction. We certainly were surprised, at that period, at finding the ground so different from what we expected, that I was willing to admit the hypothesis, and try to explain what we had seen. Looking over the entries in my journal from the 6th to the 9th of April, I see that the lands extended all round us; to the north, west, and south they were low, and were seen a very long way. Hence, admitting that we crossed Brentford Bay on the 5th, one thing still remains unexplained, viz., the land to the north; that is to say, connecting Somerset with the western lands. Captain Kennedy forgets this, and declares—1, that Somerset is an island; 2, that there is a passage between Regent's Inlet and the western sea; 3, that the sea runs without interruption from Cape Walker or Somerset Island to the Magnetic Pole. I do not think I can subscribe to this last proposition; and, as to the two others, I can only affirm them as probabilities. This does not satisfy him; but I am fully resolved, happen what may, never to support, by my assent, what I am not sure of. We differ also as to some details; for the route traversed on the 6th, which I believe to have been in two lines at an angle, he now affirms to have been in a right line, but westward, instead of S.S.W. and W.N.W., as the compass showed me. He trusts to his memory for hours, directions, distances, and the various details of our journey; and is not pleased that I will admit nothing on the authority of his memory or my own, after a lapse of two months; and that for details of this nature I prefer relying on the journal I kept regularly, and wrote up every evening. I know not what can be his reasons, unless it be that he is now afraid of finding himself in contradiction with Sir James Ross. Very fortunately for me, there are unanswerable proofs that at all events he waited rather long before he changed his opinion; and of these proofs the best are—

1, that we moved westward when we ought to have gone south, had he believed at that time that this was possible; 2, the name of Grinnel *Inlet*, which shows the persuasion he was under up to Cape Walker, and further, that the passage was closed; 3, the notice he left at Port Leopold. What confirms me in my opinion is the report of the doctor; whose plain—so wide in both directions, north and south, east and west—is, according to Captain Kennedy himself, the sea, and was bounded by the lands I saw to the north.

These circumstances make me very unhappy, because Captain Kennedy does not understand to what a degree I am bound to be scrupulously accurate in these matters, at the risk, and I may truly say, at the cost, of my peace and quiet. I know not what may hereafter be thought of the contradiction I thus set up to what Sir James Ross supposes he had seen: some will see in it the pride of a presumptuous young man, who thinks himself competent to assail the judgment of a veteran of talent; others, perhaps, will suppose I am actuated by a feeling of offended national vanity, when, God knows, I have had almost daily to do battle in defence of the performances, not only of the Rosses, but of Parry and others. I unfortunately consented to make some indispensable observations, and people will naturally consider me as having had charge of these observations, imagining that I was appointed to that duty. Thus, from all from which I hoped to derive some honour by dint of devotedness and zeal, I shall gain, perhaps, nothing but mortifications and disgusts. Alas! all the blossoms of spring do not yield fruit in autumn.

10*th June*.—To return to the same subject as yesterday: the observations I have taken here show me that there is no reckoning on the pocket chronometer we had with us, consequently our longitudes can only be estimated; otherwise there would have been an arbitrator between our differences of opinion. Altogether we have suffered more, travelled more, and endured more privations than any previous expedition; but our labour has produced no results.

But who could call me to account for the success of an expedition in which all that could be expected of me was, that I should zealously execute our commander's orders? I have therefore determined to leave each man to his own responsibility, and to be very careful, on our return to England, neither to state nor write anything of which I am not perfectly certain—washing my hands of the rest. When I look back to the early days of our voyage, and the hopes I then conceived, how different they appear from what the reality has been! Hope is a waking dream, men say. Yes, but it is a dream that never occurs twice.

14*th June*.—We had snow on the 11th, and again this morning. The seams of the deck have been opened by the winter cold, just as they are by the heat of the tropics; and yesterday a shower of wet snow, very much like rain, almost converted our cabin into a *douche* bath. Last week was employed in trenching the ballast, retrenching the hold, and in passing the running rigging. The sails have been overhauled to dry, and they have kept in as good condition on the yards as in the driest sailroom. It is worthy of remark, that the first navigators who wintered here thought it necessary to unbend the sails. The ship has not made more water during, or in consequence of, the winter.

One-half only of the doctor's twelve patients still remain seriously affected; the other half are in a fair way of recovery. A fresh gale has been blowing from the east during the last three days—a thing we do not now complain of, for gales of wind can alone break up the ice in which we are imprisoned, and drive it from the coast. The easterly gale and the swell will do the first, and the westerly gales the second.

This evening, for the first time this year, we have had real rain—that is, water which passed through the atmosphere without freezing.

15th June.—I proposed to Mr. Kennedy that he and I should repair with one sledge and the dogs to Fury Beach and Brentford Bay, in order to solve the painful doubts which our discussion has raised; but he does not think it possible to do so, and expects that we shall be able to send a boat there, or at least as far as Cresswell Bay, by searching which he thinks he can decide upon the rest.

I very much fear that no boat will ever be sent in that direction, unless it be to Fury Beach, where we have left a cask full of things, books, &c. I do not even know how we can do that, as we have but one boat now on board the ship.

17th June.—The fine weather and the flood of light in which the bay is bathed, would make us forget that we are in 73° of north latitude, if it were not for the flocks of geese and ducks, which, as they fly northwards, remind us that we alone are kept back; but, to keep up our patience, we are repairing the implements with which we made a passage for ourselves in Baffin's Bay, such as long saws, destructive crackers, &c.

The snow, which is still tolerably compact in certain places, forms a regular reflector; so that in one direction it is impossible to make a long excursion in the daytime without returning snow-blinded, and in the other the heat is as intense—I might almost say as unbearable—as that of our burning summers. It is difficult to persuade one's-self that the solar rays do not emit more heat (and have even less direct influence) than at the equator.

We have availed ourselves of this fine weather to paint the vessel and caulk the deck. The topgallant-sails are squared.

21st June.—Captain Leask has killed, at two different times, five geese and a partridge on the flat marshy land at the bottom of the bay, and two hares have been seen; but these animals are very wild, and will not allow any one to approach them. Almost all the birds we see fly in couples; those that have been killed weighed, on an average, three pounds, and their stomachs contained grass and sand. They evidently feed on the buds of a kind of heath, with which the shores of Batty's Bay are covered, and the pretty pink blossoms of which remind us of happier regions: but even here spring claims her dues; and although scentless, these tufts of heath, together with green and reddish-hued lichens, form a very pleasing garden. As we found these plants in February, and even under the snow in winter, it is evident that it is under the snow itself, and in the species of hot-house formed by the sheet of ice, which the radiated heat of the plant curves into a dome above it, that it finds shelter from the external cold. I must leave it to botanists to determine what moisture these plants find in this earth, so stony, or frozen so as to resemble stone. The driest parts of the soil are covered with the branches of the dwarf willow, the shoots of which, from fifteen to thirty *centimètres* long, creep along the ground or spread out like fans. Can this plant be really considered a shrub?

Every sinuosity of the ground serves as a drain at the melting of the snow; and from every part of the bay, at the foot of every ravine, run streams—torrents in miniature—foaming and noisy, sometimes falling in cataracts over fragments of calcareous stone, sometimes disappearing under the thickest layers of snow, to reappear further along the shore, where they spread out in sheets, and with the rain contribute to hasten the dissolution of the bay ice. The water that we drink was for some days collected on the surface of the floe; but either by filtration, or rather from the effect of capillary attraction, the sea-water, having risen to the surface, has made it brackish, and we now roll our water-casks on shore. It is also more convenient for our men to wash in these streams, and spread the linen out on the flat stones, where the sun and the warmth of these stones dry it very quickly. I shall perhaps surprise more than one sceptic by saying that a day at this season presents delightful scenes; but does not our mind grow in unison with what surrounds it? and because autumn scenery does not produce upon us impressions of the same kind as a flowery spring, or the *Moissonneurs* of Leopold Robert, is its beauty ever denied? The birds do not sing, it is true, but the beauty of their plumage makes us forget that, especially when, in the silence of the bay, at the hour when the labours of the day leave our men time to smoke, we listen to the almost harmonious murmur of the snow as it melts, and hastens from every part to accelerate the moment of our departure. On the 19th, we heard a noise like the roaring of thunder, but to which we are now too much used to mistake: it was caused by the falling of the snows in the ravines, or of earth loosened by the filtration of the water, as well as by excessive cold or heat.

26th June.—Messrs. Kennedy and Smith bring back, as the result of their sport, two eider-ducks and a partridge; the male has a yellow excrescence on his nose, not usual in the eider-duck. They saw three cranes. As yet only one seal has been killed—not that it is difficult to get near them, for, if you walk up to them and whistle, they take very little notice; but what keeps them at a distance is the pursuit of our dogs, who seem to take pleasure in driving them from hole to hole, or along the cracks in the ice, where we see them in great numbers. The other day we counted fifty in one flock. Their food, no doubt, chiefly consists of the little shrimps I mentioned last autumn, and of a kind of whelk, which we discovered on Thursday on a fox's head which was sunk in the sea. These whelks, rather more than three *centimètres* long, are capital eating; but what made them especially interesting to us was, that until that moment we had been unable to account for the presence of bivalve shells, which we had found even at an elevation of fifty feet in the bay. The floe is covered with large leaves of alga marina (some are fifty *centimètres* long, and thirty or forty wide), which come up to the surface whereever the hole of a seal exists, or even through the fissures of the floe. These plants, discoloured, doubtless, by their growth without light in the obscure depths of the sea, are not the only sea-weeds found in these regions, for we saw tufts of wrack floating in Baffin's Bay. Light is not, however, indispensable to the colouring of plants, for it appears that certain green plants grow in mines. Some leaves which I have since seen were, as well as their stems, of a very pale yellow, the leaves being sometimes shaded with the colour of wine lees.

We saw a bear on Saturday evening; he proceeded southwards, although we burned fat to attract him.

3rd July.—During the last two days we have had an easterly gale, which reminds us of winter storms; but we rejoice at the thought of the havoc it will cause among the ice of the strait. At the beginning of the week, Mr. Kennedy perceived that water now extends along this coast as far as Elwin Bay, which we passed on the 10th of June. We have no reason to complain, and yet I am impatient. I should like to find some active occupation which would divert my thoughts, and wish we could go by land to the place where the two boats are; but I neither can nor will take the initiative of this proposal, and I must therefore wait.

Mr. Leask has killed two more eiderducks, a male and female, for they are always in couples; the duck has a spotted plumage, exactly similar to that of the partridges of La Plata, reddish, with black stripes; the belly and crest of the drake are of a bluish black, the breast white, and on the neck are splendid greenish shades, blending as they approach the back.

Three more cranes have been seen; although far from being as savory as the brent geese and weavies, these birds are by no means unpleasant eating, if the skin and fatty parts are carefully cut off. Sir John Ross appears to be of opinion that most seabirds are edible when prepared in this way.

8*th July.*—Continuation of the easterly gale, of which the results are not so favourable as we might wish, perhaps because the wind, still blowing in the same direction, drives the ice, and retains it close to the coast. Seen through a telescope from the top of the hills, the east coast of the inlet appears free, even south of Port Bowen; and our desire is such that we are uneasy at not seeing the ice opening before us. Yesterday Mr. Kennedy tried three or four blasting-cylinders, containing from seven to ten pounds of powder, at the north point, but without success, the floes being five feet ten inches thick. Several brace of hares have been seen in front of us, but our sportsmen were unable to approach them.

On the other hand, we have been visited by several bears, one of whom came pretty close to the vessel on Tuesday morning, and stood a volley of eight or ten muskets with great coolness. We like to think that he was too far off for us. In the afternoon, a second having made its appearance at the entrance of the bay, we all went round to the north point (by the side of the vessel), and there, lying on the beach, awaited his passing near us; but either he suspected something — which is scarcely probable, since he was to windward of us—or from some caprice of his own, he disappeared in such a manner that we could not take aim at him without showing ourselves, to do which would be to lose any chance of success in case he should pass by us again. Like all we have seen, he was in quest of food, and walked along the clefts, into which he plunged every now and then; reappearing on the ice, he rolled in the snow, his legs in the air, grunting with pleasure, like a donkey in a dusty road. When he got further on in the bay, three men left the ship, in order to drive him back upon us, and for some time we hoped that we had caught him; at last, the animal, ignorant of the nature of the bipeds pursuing him, lay down on the snow to reconnoitre; but when, after many turnings, they came close up to him, and opened their fire, he galloped off with an agility which surprises me more and more, and quickly climbing the hillocks, over which we could not follow him, was soon out of reach. One of our young dogs pursued him, biting his heels from time to time; for we saw him turn suddenly round, and give a blow with his paw, which the little beast avoided very adroitly. We returned on board somewhat crestfallen. If we could throw off our dogs after one, I think they would keep it at bay, and so give us time to approach; but they are too busy elsewhere this spring season. Two other bears appeared at the entrance of the bay in the course of the morning, but did not enter it. Although it is impossible to affirm with certainty that they are not the same, it is scarcely probable that they are so; and we think that there is a general emigration of the bears from the bottom of the inlet to the north, where the absence of ice promises them easier fishing. We continue to prepare our ice-breaking instruments, sharpening saws, &c. &c.

10*th July.*—On Friday, having gone to the north point to take angles, I saw another large bear close to me; but on discovering me he altered his road to a southern one, although without showing much alarm. Once on board, I followed his movements with the telescope; when he got close to the shore, a falling down of snow took place close to him, without his noticing it. It is curious, that whenever these animals have approached the ship, the least noise has made them prick up their ears; they scent the air anxiously, and keep continually on their guard; but the voices of Nature, to us so awful, do not affect them in the least. Mr. Kennedy, in speaking of this, told me, that in the forests of Rupert's Land he has often noticed similar phenomena among the moose deer, which are among the most distrustful of animals; they are perfectly at ease whilst the storm roars through the branches and uproots trees; but if, in the midst of the hurly-burly, they hear some sound proceeding from another cause, such as a branch purposely broken by a Canadian, they hurry away instantly.

Yesterday we tried to cut through the ice close to the ship, in order to throw in a net; it was five feet ten inches thick. One of our exploding cylinders has been tried, but again without success, although the shock shook the schooner, the nearest part of which was twenty yards off.

The ship is still supported by the ice, so that it is probable that during the winter it had little or no water under its keel. Two of our men, who visited the hills last Thursday, report that the tableland is everywhere covered with snow, just as it was last September, and so deep that they sometimes sank into it up to the waist. It is curious that during the whole winter we had but very few days of real snow, the drift having been the characteristic of the season.

I found the small rivulets in the ravines considerably enlarged, either by the rain of the last few days, or the increase of melted snow.

The steep cliffs that rise up at both points of the mouth of the bay are alone naked, and their hideous sterility contrasts strongly with the bay. As I have before mentioned, these rocks are formed of layers of primary calcarious stone, generally horizontal, sometimes placed regularly one upon the other, but in other places piled together in the strangest way, so as to form craggy boulders.

12*th July.* — Two more eider-ducks (a couple) have been shot. Mr. Kennedy went this morning beyond the bay and the summit of the north cliffs; he has seen water along the coast, extending to the other side, and to the south, about two miles north of the ship. There are some gaps between the north and south points of the bay; but what amazes me is, that there is no water now in spots where there was water during the whole winter. Last Saturday, the ice extended along this shore as far as Elwin Bay, and this clearance took place during the gale of Sunday. A day or two will do a great deal. On his return, he found at the foot of the cliffs, amidst the fragments of the avalanche I have spoken of, a layer of a kind of salt, unknown to us all, about an inch thick, which seems to have been brought from the top of the rocks by the snow; the salt, deposited in tablets, probably exists on the ground in a state of efflorescence.

To-day the body of one of our dogs, that died six months ago, was found under the snow, and when cut up for fishing baits, was found perfectly fresh. We now devote ourselves to the whelk or buccinum fishery, on a grand scale; they are capital eating, and half a hundred make a good dish for our party of seven. I imagine them to be excellent in the spring for scorbutic patients, in default of fresh fish. Mr. Kennedy finds a letter of Lieutenant Robinson, which he did not remember to have received. It appears that the deposit of provisions south of Cape Seppings, which I looked for so anxiously in October, is in the ravine itself in which we encamped, or at least was deposited there in 1849.

The pink-blossoned heather is already faded, and almost destroyed by the high winds; and we now find some small yellow flowers, of which, owing to my botanical ignorance, I do not know the name. I am a little more fortunate, although not less ignorant, in conchology, for I recognised three different species among the shells found on shore (all three bivalves); a species of *cyclostoma*, found on the ice, and a kind of muscle, besides the *buccinum*, which the net procures us. I also found among the marine plants the fragments of another shell, or of a zoophyte, with which I am not as yet acquainted; and this evening one of our men brought me a fine caterpillar from the land.

How many things I should prepare, and what interesting studies I would undertake, if I had to begin such an expedition over again!

Captain Leask killed a male hare this evening, close to the spot where he mortally wounded the female a few days ago. The hare has some tufts of grey fur. Messrs. Kennedy and Smith shot on the surface of the water twenty dovekeys, two gulls, and a kittiwake.

13th and 14*th July.*—This morning I went with Mr. Smith to the spot where the open water begins, and in the course of half a day killed fourteen dovekeys and four gulls. (We had with us the Halkett boat, the use of which I cannot too strongly recommend under such circumstances; it would be better if the inflated portion were divided into compartments, in case of accident.)

At first our sport was plentiful; hundreds of dovekeys—lovely birds with black bodies, and white spots on each wing, and coral-red feet and beaks; numerous flights of ducks—eider-ducks, wild-ducks, and some geese—passed not far from us, but still out of reach; we also saw a white whale, and some young seals of this year, too timid to give us the least chance of killing one. One of us kept on the ice, whilst the other, paddling the boat along in the Esquimau fashion, got out to sea, drove the game on to land, or picked up all that fell. Nothing is more amusing than to see the young seals, like little monkeys, poking out of the water a tiny head, which disappears at the least alarm with a loud snort. They must have an instinctive knowledge of the animals inimical to their race; for, whilst the screechings of tribes of kittiwakes, and the loud noise made by the white whales as they throw up columns of water and mucilage, do not alarm them, the sight of a hat peering above an iceberg instantly causes them to disappear. As to the dovekeys, which we shoot most of, because they are the best eating, they dive with such rapidity, either at the sound of the cap or at the flash which precedes the detonation, that we have often seen (apart from a sportsman's vanity) the shot hit all round the place they were in; and the little birds reappear at some distance, often after a lapse of thirty seconds or more, gracefully poising themselves, and dipping their pretty little heads under water. After an hour's shooting they grew rather less daring; but as soon as we relaxed in our pursuit, they flocked in tribes to the edges of the ice, along which they no doubt find their food, in the shape of insects, seals' dung, shells, &c. &c. About twelve, the heat of the sun having become very great, cracking noises are heard in the cliff, and the stones roll down the inclined planes which lie below the horizontal layers of rock; each little stone, falling on the thin slate-like fragments, makes them resound like so many potsherds. Much ado about nothing.

The height of the cliffs is so great that the birds make a prodigious noise as they fly down upon the ice. I do not know whether this is to be attributed to the resonant property of this atmosphere.

In short, either because it was a long while since I had taken any exercise, or because it made a truce in my melancholy thoughts, I found myself all day deeply interested by the sight of the sea and its millions of inhabitants, *sub et super* (below and above), by the contemplation of the various mosses and grasses, or small plants, which peep through the stones wherever the snow leaves any moisture, and I return joyously on board; all the more joyously, because I hear that the boat is to go to Port Leopold to-morrow, in order to bring back our two other boats.

The captain brings back from his day's shooting three young entirely grey leverets, the offspring of the parents he had killed. I am told that the true European hare never has more than two at a birth; and it has always seemed to me that the Arctic hare rather resembles the rabbit than the hare.

CHAPTER XI.

VOYAGE TO PORT LEOPOLD.

From July the 15*th to the* 21*st.*—I AM very glad that I asked Mr. Kennedy to allow me to form part of this little voyage, in order that nothing may occur in which I have not my share of fatigue and labour, even when of the least interesting kind. Ten men and the dogs drag the *mahogany* to the opening with considerable difficulty, although we have nothing but the masts and oars, with the sails; and I doubt whether the boat's crew, composed of five men only, could in any case suffice to the urgency of a moment of danger. At six o'clock we set off once more, with our bedding and provisions on a sledge. For the first time, we see among the other winged gentry the loons, which differ from the dovekeys in this respect, that the latter have only one white spot on each wing, whilst the loon has a white belly, and is generally a larger bird. With one discharge of his gun, Mr. Kennedy killed nine, whilst we were unloading the sledge: they are not such agreeable food as the dovekey, and must be freed from all fat before cooking.

The ice is being constantly undermined by the incessant action of the surge, which breaks away a small portion of it each time; and Mr. Anderson, who had gone a little too near the edge, felt the ice give way under his feet: he fell into the sea, fortunately within reach of the boat, and escaped without anything worse than a cold bath.

At the moment of starting a thick fog rose, which hid from us both land and the road we wish to take, because it was driven by a slight southerly breeze. This is the first real fog we have observed, either because it never exists in the harbour on account of the absence of water, or because we went to bed too early to see it; yet during the last month, between five and six o'clock, the vapours rising from the earth always formed a sufficiently thick mist to veil the sun and hinder my observations. I picked up several sea-urchins on the floe, the largest of which was four *centimètres* in diameter; they are dead: their shells have, doubtless, been carried away by the grasses which at present are strewed over the floe, and which I erroneously supposed came from the bottom, whereas they had been evidently retained and incrusted by the ice in September, at the time of the formation of this same floe.

The noise of our course drives away hundreds of volatiles of every species, cursing in their clamorous voices, and in every tone peculiar to sea-birds (Toussenel would not fail to find an analogy in them with the harsh, hoarse accents of our blue-jackets). Several white whales came round our boat, and when at a certain depth emit a sound which sailors call their whistling: it is, doubtless, caused by the use of their organs of respiration. Whales are very common in Hudson's Bay, where they are easily caught, on account of their white colour, which enables them to be seen at a great depth. The whalers paint their boats white that they may not alarm them. Although timid when aware of the existence of danger or pursuit, this cetaceous animal appears to be of an inquisitive disposition; for both to-day and on the various occasions when we have met them, they have always passed close by us; and last Tuesday Captain Leask, in the Halkett boat, was forced to land, fearing that two of these whales, who were pursuing him, would upset the boat. Sailing and rowing by turns brings us, at four in the morning, to the spot where our two boats are (they have not been disturbed since the 3rd of June), and we re-embark after taking a hot and comfortable cup of tea.

On the cliffs which precede Cape Seppings, we recognise several slender whitish veins of the substance brought on board; and the channels which descend to the sea-shore serve as ducts for masses of water and snow, which descend from the heights with a noise certainly more formidable than the volume of the bodies causing the tumult: under the scorching sun which precedes noon, these detonations follow, without interval, from every one of the thousand channels of the coast. The cold, damp night air is succeeded by almost unbearable heat; there are scarcely any icebergs before us; and, if it were not for the torrents of snow carried along by the inexhaustible springs of the coast, we should believe ourselves to be some sixty degrees of latitude further south.

Visiting these spots naturally recals our reminiscences of last September and October, and the contrast can awaken in us none but feelings of joy and pleasure. Towards noon, on Friday, the 16th, we land at Whaler Point; about a quarter of the length of the bay is free from ice, and the Point, now perfectly bare, displays to our enchanted eyes the abundant deposit of provisions, of which we have hitherto only seen a part.

Our splendid snow-castle of the month of May had of course vanished, carrying with its ruins, and in singular promiscuousness, fragments of provisions, of old shoes, and of old newspapers—so avidly read in our hours of scorbutic idleness—our woollen roofing, and tools covered with rust. This confusion was easily repaired, and we soon constructed a tent with what remained of the materials left by Sir James Ross. Letters, addressed to the unfortunate crews of the *Erebus* and

Terror, were lying on the shore, more or less injured by damp and the sacrilegious teeth of foxes; and it was not without a pang that we picked up the fragments, wrapped them up with care, and put them in a place of safety. The various parts of the engine were scattered in all directions, and incontestably useless; but there was no room to doubt the care with which everything had been deposited on shore; only it is probable that icebergs, carried there by high tides or dashed there by the sea, must have borne them away. I likewise think that this point, composed of a friable soil and small stones, is incessantly worn away by the action of the waves. The long-boat is still in good condition, for the water formed by the melting of the snow had not escaped from it. We spent Saturday in arranging and hauling a little further on shore its various appendages, including an anchor weighing three hundred and thirty pounds, and a strong machine for launching it. We could not attempt to launch it—our strength would not have sufficed; but it is to be feared that it will soon be carried away by the ice. It was evident that it had been placed where it lay only to render possible its being easily launched, even by a party reduced in strength and number. All the barrels have been laid on their sides, that the water might not lodge on their heads and soak through them; the sails are carefully folded and hung up, along with needles, canvas, twine, &c., to repair them; and all the species of awning we could find. I think that, in an expedition like that of Sir J. Ross, it would be better to build a certain number of small tents than one very large one, on account of the difficulty of heating the latter, taking the chances of wintering into consideration. On Saturday evening a bear crossed the bay, but made rapidly for the south, after the wind had made our presence known to him. He was, no doubt, one of those to which we had already given chace.

On Sunday I took a walk to the graves at the bottom of the bay, accompanied by Kenneth, the carpenter. The snow forms round this part a chain of small fresh-water lakes, on one of which we thought we perceived a crane. The head-boards of the six graves, dug by the *Enterprise* and *Investigator*, had been blown down, and we set them up again with the solemn feelings which such a scene could not fail to inspire. All the graves were those of young men between the ages of twenty and thirty. We visited the side-stone, where these marks are cut on the granite—

$$\left[\begin{array}{cc} \underline{\underline{E}} & \underline{\underline{I}} \\ \multicolumn{2}{c}{1849} \end{array} \right]$$

Facing us, to the south, is a ravine, in which the snow still lies in beds thirty or forty feet thick. The mosses and heaths are still in bloom here; at Batty Bay they are already withered and dried up: it is, no doubt, in consequence of the difference of latitude that they blossom here a few days later, for Point Whaler is swept by all the winds that blow. The weather was foggy by day, and in the evening a strong breeze got up, accompanied by rain.

On Monday morning we find that the ice from Lancaster Sound has filled the bay, and the coast is encompassed by a bank of hummocks wider than the ice was in May. The surge heaved these hummocks, and made them clash together in a way by no means desirable for a boat that should venture among them, so that we must postpone our intended return. Fine weather reappeared on Tuesday, the 20th, but without bringing about any change in the ice around us, which does nothing but circulate round the bay with the tide.

One of our men thinks he has seen a vessel to eastward, and we all go to the foot of Cape Clarence to have a better point of view; but we are obliged to give up the hope so quickly conceived of news of the other expeditions, and of letters from home. What must have been the tortures of the two Rosses in 1833, when, after an absence of five years, they watched on the same spot for signs of an opening in the ice, or for the appearance of a ship, which alone could put an end to their miserable condition! The point is strewed with bones of whales, and we counted no fewer than twelve skulls of those animals—a circumstance to which, no doubt, it owes its name. Like the eastern coast of the bay, it contains also numerous ruins of Esquimau winter dwellings, with thick walls, and subterraneous communications from one to another. We searched, without success, several of their graves, in which only some bones had escaped the voracity of the foxes and bears, or the somewhat sacrilegious curiosity of the different visitors. These graves are not sunk in the soil; the body is generally laid on the surface, or in a shallow excavation, with the principal fishing utensils of the deceased, and covered over with stones, between which wide spaces are left; "because otherwise," say the Esquimaux, "the dead could not take their flight to the eternal abode." I took part, like the rest, in these searches, hoping to find something that might be interesting for ethnologists; but I have often thought since of the significant manner in which Adamson shook his head, and told me that it would be better to leave the dead at rest. Among ourselves, where certain statistical measures furnish science with the elements of which it has need, the violation of the grave is an unjustifiable profanation; but I cannot for a moment believe that the least blame can attach itself to such researches, when they are not undertaken wantonly or without a useful purpose. In

Montevideo I have galloped over saladiros strewn with carcases of oxen; but my horse reared and threatened to throw me when we passed near the carcase of a horse. Oxen often refuse to enter a slaughter-house; here our dogs feed on one of their companions that has fallen on the ice. Is this only a characteristic difference between carnivorous and herbivorous animals?

21st July.—The ice, without having quite disappeared, yields us a passage, of which we hasten to avail ourselves; for it is easy to account for the observation of Sir James Ross, who in 1848 entered the bay on a certain day, and could not have done so at any other period of the year. From six o'clock till noon a light breeze from N.E. carries us to the second ravine after Cape Seppings, where our two boats lie; and there we find also the masts and the sweeps of the gutta percha. This is the ravine that so cruelly disappointed us on the morning of the 10th of September. We quickly repair the gutta percha, and make it stanch with bands of stuff prepared with naphtha, &c. This facility of repair is not the least valuable quality of this very light kind of boat, which has the further advantage of not being liable, like others, to be injured by the ice. The mahogany boat and the gutta ship five sweeps each; and on our return, when we had to transport our provisions, the same number of men was necessary for the former, carrying sweeps and sails, and 190 pounds of flour; and for the latter, which carried, along with its sweeps, three quarters of flour (570 pounds), two of preserved fruits (90 pounds), fifteen gallons of lime-juice (151 pounds), two barrels of herrings (218 pounds), 130 pounds of salt pork, sixty-five cases of preserved meat and vegetables (455 pounds); in all, 1600 pounds, without reckoning four muskets; making a difference of more than 1000 pounds. I think, then, there is not a shadow of a doubt as to the advantage of this kind of boat in moderate weather; but this advantage is annulled in extreme cold, for then the gutta percha becomes brittle.

Elwin Bay, which we pass at a distance of two or three miles, appears still blocked up with ice. Steering between floating blocks, or through the streams which are not compact enough to bar our way, and the pieces of which we push asunder, we glide slowly along the coast, again followed, or even pursued, by white whales; and at two in the morning we are a little further south than the spot where we launched the boat on Thursday last. The eastern ices are in some places closer together than they were at the same date, in consequence of the gale from the east; but we see with pleasure that a large portion of the floe which closes the entrance of the bay has been detached, and a few blasts of wind from the west will soon clear away the rest. After landing our cargo and putting our three boats in safety, we reached the vessel again at four A.M. on Thursday, the 22nd.

CHAPTER XII.

ON BOARD.

Nothing new on board. The captain has killed a large seal, a clapmatch, nine feet and and a half long. He had to hold it back by one of the hind fins, shouting for help to the ship, and fearing every moment to see the huge animal vanish from his grasp.

24th July.—The last three days have been spent in carrying our boats and provisions to the vessel, though it is but three miles from the latter to the place where they all lay; but it is now rather difficult to cross the ice, on account the pools that cover it in many places, and which are rendered very dangerous by the holes made by the seals, because these are not easily seen under the water; and a man dropping through one of them could not perhaps hit the opening again, and come to the surface, unless he was a good swimmer, and not too soon overcome by the cold. Two of our men fell into the water this morning near the vessel, and the current, which was carrying them under, might have made their position very critical, but for the prompt assistance they received.

In our several trips these last three days, we have observed that the main clefts are widening, and to-day it was not possible to land at the place to which we used to go so easily a week ago; but it is to be hoped that the warnings thus given will prevent all accidents. Several symptoms make us uneasy with respect to the ship's getting out—not as to our final liberation, thank God! —but at least as to the question, what shall we have time to do before we think at last of steering homewards? Yesterday evening ice was forming on the surface of the pools round us; it acquired a thickness of several lines during the night. At noon to-day, with the thermometer at 57°, it had not disappeared.

26th July.—It has been resolved to have recourse to grand measures, and the saw has been set to work to aid nature in effecting our deliverance. During the last month the vessel has become inclined two feet to larboard. It is probable that on that side on which the snow accumulated most easily, on account of the shelter of the vessel, and on which we had placed our staircase, carpenter's shop, &c., the ice is thicker, and that, pressing the vessel below the centre of gravity, it forces it to lean to larboard; for there is more water, and the ice is more decayed to larboard than to starboard. Yesterday the tin skiff filled and sank there. The ice having been sawed all along the starboard and abaft, the vessel righted a little, and in the afternoon completely recovered its equilibrium. My reasoning of this morning is confirmed by the fact, that the starboard side, which was under water, and on which the vessel is now leaning, is quite dry.

The floe appears to be detached from the land all round the bay, and when the entrance shall once have been cleared, a strong breeze from the west will suffice. I hope to sweep all remaining obstacles away. Mr. Leask, however, is afraid lest the icebergs grounded on the bar, and which were there even in September, when the bay was entirely clear, may retain the floe for some time, unless it passes away piecemeal.

Already in idea I transport myself to Pond Bay, but there will be no whalers there at this season to give us news. The chances are not in their favour when the land-floe has disappeared, because it is on its margin that they find the whales. The ice appears to be wearing away at the under part of the floe, and not at the upper part. The heaps of rubbish we have thrown out during the winter are not yet sunk, and they have preserved the ice by their thick volume; whereas we observe that wherever they are spread in a layer, they sink readily. Pieces of grey limestone which we throw on the ice make holes of no great depth: it is the most highly coloured sea-wracks that disappear fastest, some of them sinking to a depth of two or three feet. It strikes me that, by loading the detached pieces of ice with sand or black gravel, they might be sent to the bottom, or easily pushed under the adjacent blocks; but then the saw would strike upon the gravel and be spoiled. We stretched a line from the after part of the vessel in the direction of the northern cairn, and cut and sawed the ice under it. This job occupied eleven men for eight hours. The ice appears more rotten, as we sailors say, where it is covered with water. I think, then, that, when it can be done with safety to the men, the line of sawing should be made to run as much as possible across the pools. Good boots would preserve the men from the risk of getting wet feet.

Our invalids and convalescents are going on well. Miller and Magnus are the only two of the ship's company not quite free from all symptoms of scurvy, but they can take part in every kind of work, even outside the vessel. The doctor went ashore two days ago; so there remains only poor Mr. Hepburn, whose worst complaint is incurable, as he himself jocularly remarks, for it consists in being twenty years too old.

Took in the slack of the chain at high tide. It would evidently be better to do so at low water, in consideration of the state of the ice.

29th July.—It took not less than three days to saw the ice through from the vessel to the north point, where there is a large cleft crossing the bay southwards. The saw was damaged the first day by a stone, probably carried and held there by sea-weeds; for the ice forms by layers added from below. The crow's nest has been again set up at the mast-head, and frequently in the day we climb into it, and anxiously scan the horizon. Whatever be our hopes, it is at least greatly to be feared that we shall have but very little time left us after our liberation. It is now impossible to reach the cairn dryshod, and the men engaged in sawing are obliged to take a plank to cross the pools. The taking-in of the slack of the chain having been continued, the vessel has formed for itself a little basin, still of very narrow dimensions, and it is now supported by its natural element. Mr. Gideon Smith spoke to me yesterday of a bed of moss made by the Shetland fishermen. His description is very like that of *Polytrichum commune*, given by Linneus in his journey in Lapland. (*See* M'Gillivray's *Lives of Eminent Zoologists.*)

30th July.—Yesterday Captain Kennedy and I went ashore in chase of a hare we descried on the hills to the north. We could not have reached land at high water without the aid of my cloak-boat, though the thing is still practicable at low water. We went over the ground which seems to be the favourite haunt of those animals; but it was to no purpose, owing, no doubt, to the wise precautions of nature, which gives them at all seasons a fur of the colour of the ground, whereby they more easily escape their various enemies.* Our young pets are very well off on board, since we can supply them with fresh herbs every day. They also eat preserved carrots very readily; but what they seem to enjoy above all is the flowers of whatever plants we bring them; among others, a species of yellow ranunculus we found to-day.

It has been snowing and raining since noon. The ice formed during the night was half an inch thick. From the hills the ice

* Without entering into a speculative digression as to effects and causes, I think at least that this is the most plausible explanation of the change in the colour of the fur of Arctic animals.

appears to us to stretch to the other shore of the inlet, and from the mast-head it is seen to extend only a little northward of our bay. We can see the open water flowing in a sort of lane between the two points, as far as Fury Beach.

Our men began sawing again yesterday, in a line nearly parallel to the first; but as it passes through the pools, the work goes on more rapidly—the average thickness of the ice being three or four feet. In the ravine immediately eastward of the vessel, the torrent is five or six feet deep in many places. We could only cross the other yesterday by making bridges of stones. The low point on our side—that is to say, to the north—is fringed with ruins of encampments, more recent than those of the flat lands to the west, where are the winter residences, on a more humid and now marshy soil, because the snow is thicker there than in winter. I believe these places must be frequented by the same tribe which now inhabits Boothia, for we find its holes all along the coast.

The second line has been made to diverge gradually from the first, as it approaches the crack forming the corner, so that the mass may be more easily pushed outwards; but towards the ship the contrary has been done, in order that the first segment may withdraw inwards, there being a considerable extent of water near it. As the vessel has not room enough yet, the stern is kept fixed by an ice-anchor; and the northern chain is strained in order to haul us as close as possible under the shelter of the land, and where the floe is best, for fear the bay-floe should drag us away with it. The two lines extend to within thirty-three feet of the crack, and the inclination is a foot per fathom. The distance from the stern of the vessel to the crack is about fourteen hundred and four feet. The operation, though fatiguing and disagreeable, since our men are generally wet-footed in spite of their boots, is yet one of those of which they make no complaints, in consideration of the end proposed.

Yesterday I killed a species of plover.

31st *July*.—Yesterday afternoon the weather, which had been rainy all day, turned to snow, so that at eight o'clock I required light to read in my little nook. The thermometer had fallen to 31°, that is, one degree below the freezing-point of water, which it had not done in the open air for some time. At ten o'clock we had six inches of snow on deck, and this morning we are all surprised at the aspect of the land, which reminds us of the worst days of winter, the cliffs themselves being covered with snow on the sides of the straths, and on their buttresses, descending to the ice. It is easy to conceive the effect produced by this change: my reflections upon it lead me to consider our excursion of the day before yesterday. In many places it appeared that the rains and the filterings from the snow were washing away a large portion of the friable part of the soil. The excessive frosts of winter, and the summer suns that break down the faces of the cliffs, and the ravages which the sea and the ice-blocks occasion on the coast, must materially alter the figure of these regions in a determinate space of time; and I think that a model in relief of the topographical outlines of certain parts of the country, carefully executed, would be a very interesting acquisition to science. I believe that observations like those I mean have been made on some glaciers of the Alps.

As I wrote yesterday, the vessel is afloat, and, what is more, perfectly stanch, her pumps not bringing up more water out of the hold than before winter; neither has our rudder suffered.

The barometer has fallen from 30° 5' to 30° 2'; and the breeze, which blew east at seven yesterday evening, north at eight this morning, blows north-east at ten this evening. Last Thursday we at last found on shore a plant which seemed to us to resemble sorrel; but it is quite stunted, for its leaves are barely an inch long and a few lines wide, though the plant is at its full growth.

The fallen snow, which now covers the ice in the shape of fresh water, will doubtless contribute to the destruction of our prison. August, too, is at hand, with the gales it generally brings in these regions, and with welcome promises of deliverance. At eight o'clock the thermometer is at 36°, and the chilliness caused by the wind makes us feel the fireside delightful.

1st August.—The wind continues to blow strongly northward; and, though that is not the most favourable direction for us, it is better than a calm, on account of the swell it cannot fail to raise in the inlet. As we approach the probable term of our captivity, I feel my impatience redouble; the few moments we have to pass here cannot, alas! be employed with any advantage to the cause in which we are engaged, and the certainty of this contributes to make the time seem longer to us.

This is the second anniversary of my father's fête day on which I have been absent from home, and my wandering life will, no doubt, condemn me to miss many another. Poor father! who are so proud of your son, and in whom family affection highly surpasses all other sentiments, how I wish I could hasten the flight of that time which separates us from the accomplishment of projects so fondly cherished! How I long to be able to make you enjoy, dear parents, for the first time, those comforts which a long life of toils and hardships has kept aloof from you!

Captain Kennedy tells me that in Canada, where the rabbits are white during the winter, they remain grey when they are domesticated. Attempts have been made at Red River to employ the buffalo in ploughing or draught, but without success; he cannot tell why. Mr. Hepburn has seen attempts to tame reindeers, but it seems that they too failed. The Company, Captain Kennedy tells me, encounter, on the part of the Indians, insurmountable prejudices, which hinder them from delivering the young fawns which they might take and break-in to work. Attempts to tame otters, but only as a matter of curiosity, have succeeded. Captain Kennedy reared two, which he lost after a while, in consequence of a malady he did not know how to remedy (retention of urine). This reminds me of the precautions taken in this respect by our bitches for their young, and which hitherto I had attributed only to a natural instinct of cleanliness. Nothing should be neglected when the acclimation or domestication of animals is to be accomplished; the breeder cannot observe their habits with too much attention, in order that he may be able to render them almost maternal cares and services, and supply the place of their absent family.

2nd August.—This morning, to our great surprise and satisfaction, we find that the ice is open from the north point of the bay to the bank of reefs, or Cairn Bank, forming a passage almost sufficient for the vessel.

It was full moon on the 30th, at two P.M., and we expected a favourable effect from the corresponding spring tide; but no such thing. The spring tide must have taken place early in the day; so the influence of the moon is felt here at two days' interval. We are now parted from the open water only by the portion we have sawed, and which would be swept out by a light breeze from the west; the ebb, I think, can have no effect upon it on account of the bar.

I forgot to mention that in our Thursday's excursion Captain Kennedy and I observed that the ice has disappeared from the bottom of the bay, even in the narrowest coves.

The first part of the day was spent in sawing the ice crosswise, so that the blocks, being smaller, might more easily get free.

The wind shifted in the night from N. to N.N.W., and perhaps contributed to this change in the state of things; but eastward the ice seems to meet the other side of the inlet, the vacancies having, no doubt, been filled up by the late north winds. I went ashore, at Captain Kennedy's suggestion, to examine the appearance of things from the top of the cliffs; but the haze and the snow that fell obliged me to return without seeing anything. The commander, besides, had requested me to return early.

Our anchors were cast without buoys in September, and it seems that Mr. Leask talks of nothing less than leaving anchors and chains here; so that, once out of this place, we should have nothing for it but to do our best to get out of Baffin's Bay, seeing that we have only one anchor on board: neither could we think, thus unprovided, of reaching England by the north of Scotland. That is not all: by the laws of salvage, a vessel without bower-anchors is considered by pilots as being in danger of shipwreck, even though it have sustained no damage; and Mr. Hepburn tells me that this would cost not less than £500.

A movement has taken place in the ice, which I account for in the following manner:—The floe having been long broken up at its edges, the high tide could do no more in that way; but, by raising the whole above its ordinary level, it has allowed the wind to drive the floating fragments to where there was a vacancy to fill. At night we see that the ices of the inlet, which come in face of the entrance of the inlet, are of extraordinary thickness, which leads us to think that they are ice from Barrow's Strait, driven first by the west winds into Lancaster Sound, and then by the north wind into Regent's Passage. I must add, too, that the wind which we have from the north in the inlet, is generally west in Barrow's Strait, according to our experience.

Yesterday we saw a northern diver—a very long-bodied bird. I had already seen one at Port Leopold, but mistook it for a stork.

3rd August.—The snow is not melted on the floe, as we expected, and a soft ice, two inches thick, has formed on our lagunes on the surface of the floe; moreover, the coast is still speckled all over by the snow showers that have lately fallen; but at our waking we find that the entrance of the bay is now quite cleared from point to point. We set to work betimes, and Lady Franklin's friends have the pleasure of seeing our second anchor weighed without difficulty. The vessel is now held by two ice-anchors placed on the floe, but forward only; for it has a little basin, in which it floats and rolls to our great delight.

We proceed immediately to clear the canal cut last week, though it must be impossible for us to take advantage of it to-day, the ice barring our passage at the north point; but we shall not have lost a minute, and that keeps our minds engaged. The sawed ice has stuck together again, as I have already said: the sawdust (so to speak) forms a sort of cement between the cut edges, which the cold nights freeze, and we are obliged to apply the saw again everywhere. It would be more advantageous then, I think, unless the work was intended to keep the crew employed, to saw only a small number of pieces, and remove them in detail; moreover, our pieces were cut too narrow towards the opening, the inclination having been, as I have said, two inches per fathom: it ought to be six inches. The job, therefore, has not been very rapidly executed; for this evening, after ten hours' work, we have cleared only thirty fathoms, which we have driven out to sea in pieces five or six yards' square, either by means of our blasting-cylinders, or by jumping on the ice all together in regular time, or by running from one end to the other, using our handspikes for levers, &c. Ice may also be broken by rolling a boat on it; but it must be very thin for that. Our cylinders acted perfectly well, except the small ones, of two, three, and four pounds; but the large ones, of seven or eight pounds, produced the most satisfactory effects.

The result depends also on the manner of using these instruments. After borin through the ice, the cylinder must be sunk, and the hole filled with snow and stones as hermetically as possible. To the end of the cylinder a string must be attached, which must be pulled, when the cylinder is under the ice, in such a manner as to give it a horizontal position, and thus expose the largest extent possible to the action of the powder; then the match is lighted, which is placed in a gutta-percha tube. I am speaking upon the supposition that the resistance of the ice is still too great to be effectively overcome by a small quantity of powder.

The sea-birds had already taken possession, which is not to be wondered at, for they always remain on the verge of the open water. Some white whales also came near us. The Hudson Bay men say they feed on salmon, like the seal, which is proscribed in Scotland for that reason. We see several flocks of black ducks. I got several pretty specimens of shells out of the mud brought up by our anchor from ten fathoms' depth.

Part of the larboard chain was encrusted yesterday in ice, at a foot from the under surface. It is probable, then, that ice acts as a suspension on chains. Parry weighed his anchors before winter. I am spent with fatigue to-night, but my heart is light, and I am going to sleep well. Hoping is living, and living is hoping.

4th August.—The same strong breeze from the north this morning. The offing appears clear of ice to a great distance eastward; but the ice forms a segment of a circle, the ends of which abut on the two heads of the bay. In the evening a strong breeze, between N.W. and W.N.W., drives these floating pieces out to sea. Continued our work of yesterday. We greatly regret we have no hand-saw, as the Americans had; they would be excellent for cutting off corners of ice, for which we are obliged to use shears.

A walrus, the first we have seen, came and played at the opening of our canal; white whales, too, came yesterday in troops to reconnoitre its vicinity. One of them was accompanied by a little blackish calf, which dived and swam all round her, and seemed intent on imitating all its mother's movements. I am told that these cetacea are generally black in their youth. A fox spotted with grey has been seen. It appears from a report of Captain Moore, that the Exquimaux of Behring's Strait domesticate reindeer.

5th August. — Same work as yesterday. Immediately upon the north-west wind falling, the ice returns to the entrance of the bay. According to astronomical data, we ought this day to take the sun. At midnight the canal is advancing rapidly, and it is probable that to-morrow the vessel will have an open communication with the main sea: thus, by a sort of magic phenomenon, our schooner, quite imprisoned on Sunday last, is on the eve of taking flight.

6th August.—In the morning we finish the work of clearing the canal. At three in the afternoon we run, in a few minutes, before a fresh breeze from N.W. out of the bay, in which we have remained nearly eleven months. After having so long sighed for deliverance, we gladly give a farewell look to those lofty cliffs, every fissure and feature of which we know by heart, and which we shall probably never see again. Those arid rocks, those snow-clad lands, which bounded our view on almost all sides, are familiar acquaintances of ours. This dead or torpid nature is for us full of life and feeling.

Farewell, then, Batty Bay; thanks for thy hospitality, such as it was. The ravines send us puffs of wind under which the little schooner bends, and seems to make way toilsomely, as if she had lost her locomotive faculties. The ice affords us a channel about five or six yards wide along the coast.

7th August.—The same squally weather. At noon we are still south of Elwin Bay, having to make head against the wind and a strong current. We are to touch at Port Leopold, to leave notice there of our departure from Batty Bay, and go thence to Griffith Island, in order to see if any document has been left there by the other vessels.

8th August.—The weather has cleared up; but we are stopped off Boat's Ravine by a stream which bars the way before we have been able to pass to the east side, where there is a great quantity of open water: it flows south-east. We bear to the south-west, to wait near the land until the ice shall have been all driven into the offing.

At noon we are a mile to the north of Elwin Bay. We were not able to see if the sun set at midnight. In the afternoon, ice is drifted against us by the wind, which is, no doubt, east in the offing, and we take refuge in Elwin Bay. Some grounded icebergs seem to indicate a bar. We hug the northern shore; it is nearly the time of high water, and the boat finds two fathoms within the bar. Beyond a low point which projects southwards from the northern coast is a bay, which the boatswain says is as large as Batty Bay, but without water. Snow and wind, varying from N. to N.W. during the night.

9th August.—We tried to get out of the bay with a light variable breeze from the north. About a mile from our moorings the captain perceived from the mast-head that further to sea the wind is east, and we made all haste back to Elwin Bay, towed by our two boats. The ice follows at our heels, and as it overspreads the bay we haul close to the north shore. The schooner is aground in six feet of water at low tide, her draught being 8 feet 8 inches. The northern point shelters us a little; nevertheless the vanguard of the floe reaches us, and makes the vessel heel. Very fortunately it is calm, and the velocity of the ice is not great, nor is that which reaches us thick: the thickest grounds at the point, and so the mass is stopped. At low water we are completely aground, and thrown on our starboard beam. What has happened to us shows how important it is always to watch the movement of the ice. Had we been two miles further north ($2\frac{1}{2}$ from the north point), as it was nearly calm, it is certain we should not have had time to tow the vessel into the bay, and it would have been crushed against the icebergs aground on the coast, and in danger of foundering in the finest weather.

It is also necessary to use great discernment in the choice of moorings. My first idea would have been to moor within the bar, in order to be protected by it; but, as that bar is on the southern coast of the bay, and the wind was driving the floe directly from the east, it was preferable to moor off the northern shore, on account of the hook formed by the north point. This morning we lost the stock of one of our anchors, because the forelock was not well fixed. We made another of wood, composed of a round bit in the hole of the stock, and two pieces of elm on each side. Our vessel's small draught of water is certainly a great advantage in such circumstances, the least thing serving it for a shelter; but it is easy to see how precarious is this navigation in ice, especially along the coast. This is one of the reasons why whalers abstain from trying the passage to the north, by Melville Bay, when there is no land-floe; for then they cannot make basins in the ice if the wind drives that of Baffin's Bay on the eastern coast, nor make fast to hold their ground if the wind blows off shore.

The sea was this morning covered with very small mollusca, such as doubtless serve as food for whales. The boat brought me two butterflies from the land, and I saw a gnat there. The seals, which seemed abundant the day of our arrival, have disappeared; but the black ducks are still very numerous.

10th August.—The ice entered the bay in the same way as it did yesterday afternoon, but this time without pressing us. At high water the vessel lifts, and heels to larboard, thus offering its side to future attacks of the ice: for this purpose we carried the two chains to larboard, and a stay tackle has been hooked on the ice. The master thinks the vessel will be better so, though he offers no explanation. It is needless to say that our rudder has been unshipped ever since the vessel has been stranded.

11*th August.*—The master went out in a boat, and discovers that the ice is pressing against Cape Sepping, leaving a channel of about two metres wide along the coast. At high water we float the vessel, and cause her to quit the bed she has formed by heaving from one side to the other.

A school of white whales, more than fifty in number, swims round the bay. One of our hares is dead: they are now extremely tame, and rush to the door of their cage to take the grass and flowers we bring them from land.

13*th August.*—About mid-day there appears to be an outlet, and we hasten to set sail, knowing how precious moments are.

Two crows fly over our heads, attracting our attention by their unpleasant cawing. One or two of our men shake their heads in a significant manner, and observe that this is the second time we set out on a Friday: another, somewhat more enlightened, says that in his time there were few captains of whalers who would have liked to commence a voyage on a Friday; "but," he adds, though not apparently quite certain on the subject, "it is probably a superstition."

After three or four hours' beating to windward, the ice forces us to return to land, and we are obliged to regain our anchorage for the fourth or fifth time. How long will this last? it is a question I dare scarcely ask myself. The fact is, the season is advancing; and, if this continue much longer, we must set out on our homeward voyage, or we run a great risk of being trapped for another winter. Pay attention, when anchoring in shallow water, that the vessel, in case of being left aground, does not settle on the anchor.

14th August.—A bear was signalled this morning, walking on the beach and sniffing our track: we saw it a few minutes after take to the water, to come and reconnoitre our vessel. A boat was launched, in which Messrs. Kennedy and Leask chased it. The two guns having missed fire just as the boat was upon the bear, it turns at bay, and tries to seize the side of the boat with its paws: the captain strikes it on the head with the butt end of the musket, and breaks the latter: the animal attempts to save itself by swimming; but the boat soon catches it up, and two bullets quickly despatch it. It did not attempt, like the bear last year, to throw off its assailants by diving. It is almost the only manner in which you can make certain of killing them; for, if the ice had been within reach, it would have quickly disappeared.

Our detention proves how important it is for us not to lose a minute; for, if on Sunday morning we had succeeded in passing between the two floes, we should not have lost all this week. In the afternoon we set sail again, under a light breeze; but about half an hour after midnight, at the moment we were passing between two floes, they reunite

—the one to the north running southward with a rapidity which would have been excessively dangerous to us if the floes had been thicker. We have scarcely time to unship our rudder; and this is a manœuvre so frequently repeated, that I would earnestly advise all captains who visit these seas to have the helm-port made as large as possible.

15th August.—Fine weather; a gentle breeze. The current has generally carried us southward; and at mid-day we are ten miles to the east of the southern point of Elwin Bay, plying to windward, or being towed in the centre of the pack-ice, whose difficulties require a practised eye to escape; not because there is any danger, when moving slowly, but the slightest check stops the ship's headway. It is the same with the boy-ice, which, even at an advanced hour of the day, may be seen round the floes, evidently formed during the night.*

* * * * * *

* There is a gap in the journal for seventeen days, which Bellot probably intended to fill up with notes made on separate slips, as he has left at this date a dozen blank pages in his register journal. It is probable that the seventeen days in question, from the 16th August to the 2nd September, were

2nd September.—Yesterday's gale has lessened, and we are now only stopped by a heavy swell. However favoured we may be hereafter, it is not the less evident that our passage will be lengthened some ten days. We ought to have passed to the east, in a line with Pond Bay, as is usually the case. The north winds, which appear to prevail at this period, press the ice on Cape Searle (Mollymake Head), and on Cape Walsingham; but, with the exception of these general principles, I believe all is chance and hazard, and that the pretended experience of the whalers was a story invented to impose on the ignorant; for in the last expedition, and in that of Sir E. Belcher, there were no ice-masters, they were at the most quarter-masters.

A poor little snow-bird, which has lost the land, doubtlessly in the last gale, takes refuge on board; and, very soon becoming used to us when it finds we do not mean it

employed in labours resembling those of the preceding days; and which must have been performed with some success, as, when the journal recommences, we find that the *Prince Albert* has emerged from Barrow and Lancaster Straits.

any harm, it hops about the deck, and picks up the crumbs of biscuit at our feet.

We get into a creek in the pack; but perceiving there is no opening to the east, we turn back. During my watch the swell threatens to drive us on the pack, and we are forced to paddle off as well as we can with the oars of our boats. A vessel of our size, and even a larger one, ought to be provided with at least three or four pairs of galley-oars. Only one boat is on board, so we could not row the ship out to sea. The whalers frequently have recourse to the following proceeding (called, in ridicule, sanctifying):—They raise the boats level with the water, lash them bow and stern, and ship the outer oars: the boats remain on the tackles, and out of the water. They state they make greater progress in this way, and require fewer men.

3rd September.—A calm. The same swell, which must result from another gale, to which we were not exposed. Numerous flocks of birds; all those we could recognise steering southwards: the mollymakes and ivory gulls alone remain in sight. We amuse ourselves by fishing mollymakes, which are caught with harpoons. The young mollies are grey, and have a smell of rancid oil; still our master likes them better than the dovekies. Mode of cooking:— remove the skin and all the fatty parts, and soak them in water.

5th September.—We continue our progress northward, following the edge of the pack, in the hope of finding an opening or passage to the east, running at intervals up creeks deep enough to lead to hopes, but where we generally have to sail back again; which, together with the slight wind and the sluggishness of our vessel, requires a great deal of time. A steamer would not experience the same difficulties: who knows, too, how this will end!

At our present elevation, the pack is composed of thick ice, which our master says must have been at least three years in the course of formation. It seems that the different inlets of Baffin's Bay must have thrown off the accumulations of previous years. It would not be surprising, therefore, if Sir E. Belcher found Wellington

Straits extraordinarily navigable; and he has certainly set out on this expedition with favourable auspices. Would to Heaven I were with him!

6th September.—This morning we have at length a lead or passage to the east; and, as far as we can see, clear water, free from ice. It came just in time, for we were begining to despair of success.

I am decidedly of opinion that there are only two ways, at this season of the year, to reach Cape Farewell, or the mouth of Lancaster Straits:—Sail east-south-east or due east from Pond Bay; or, if determined to follow the western coast as we have done, push southward as far as possible. The coast in the neighbourhood of Cape Walsingham abounds in ports—little bays—where a vessel can take refuge, and let the pack pass. In any case, if caught in the pack, the vessel drifts southward with it in a very short time. This has happened to several whalers, who, frozen in or caught in, were released a month or six weeks later—in November or the beginning of December; while the *Thomas*, frozen in in October, I believe, or at the end of Semtember, opposite the Devil's Thumb, was not liberated till the following March. Our skipper, who has been already caught once or twice, has lost his head, in some measure, and does not know on what saint to call.

It was, consequently, time for us to have our doubts settled.

7th September.—At four this morning, I was summoned to take my watch, with the agreeable news that the pack extends in front of us from north to south-west. The night was dark, and our vessel has been brought to through the fear of rendering our position worse. The thermometer has fallen to 26° (4° centigrade), and the ice which has formed this night is thick enough to arrest our progress, when daylight permits us to set sail. The reflections to which I yield during my watch are far from being agreeable; the more so, as the pack extends northwards as far as I can see from the mast-head. I regret that we did not remain with the *North Star;* at least our time would have been employed in a useful manner: but, as it would be dangerous to remain longer in the expectation, it is my intention, when our opinions are asked, to run the chance of the western coast, as it is too late to return to Lancaster Strait.

About 9 o'clock the master discovers a slack, and we force our way through it, striking a mass of ice on either side as we go, and receiving blows which do not disturb our equanimity. At last, by 10 o'clock, we are to the east of the pack.

I do not know, this time, whether it is prudent to yield to hope and confidence: so much is certain, that this morning I would gladly have consented not to reach England before December, on a guarantee of arriving by that time; consequently, I ought to feel satisfied with all that may happen to us. I allow that it is the rudest trial of that philosophy which I have studied, striving to resign myself to what I cannot prevent. As we have neither coal, provisions, nor boats—and even admitting that the vessel had escaped the dangers of the pack—it is certain that more than half the crew would have perished. As for myself, God be praised! my health is as good as anybody's. My confidence is in Him whose will we cannot prevent being accomplished, and I am convinced that my chances were as good, if not better, than those of any man on board; but this perspective of dangers, without any useful object or result, is not adapted to tempt even the most stoical.

8th September.—A fresh breeze from the north-west carries us rapidly eastward, and we recognise those immense icebergs of Melville and Disco Bays, so different in shape from those on the western coast—the latter being generally lower and longer, like flat islands detached from the glaciers formed on this coast in the ravines. In the morning we sight the land to the north of Uppernavik. Thanks to God! we are now out of danger, and we can henceforward sail with some degree of certainty.

Here Bellot's journal terminates. The remainder of the voyage probably afforded him no interesting observations. Once at Uppernavik, the return voyage would be effected with as much regularity as the outward voyage, and without doubt they touched at the same points on the coast or at the same islands. A few days later the *Prince Albert* sighted the coasts of Scotland once again; and the French lieutenant received, on his passage through England, testimony of most cordial sympathy, and frequently of touching enthusiasm. Lady Franklin, the Admiralty, and all those noble-minded persons who did not recoil from any danger or sacrifice to accomplish their work of humanity, recognised in Bellot a brother in devotion, in generosity, in courage—a compatriot, so to say, quoting the expression of an Englishman; for the world is divided into two great nations—the nation of vulgar souls, and the nation of lofty minds. Bellot belonged to the latter.

NOTES AND DOCUMENTS.*

Persons may always be found, at every period, who oppose the spirit of discovery under the specious pretext of sparing the public funds: some obey conscientious scruples, which I, however, regard as erroneous, as, in the first place, it may be said that the results obtained frequently surpass the most extravagant hopes; others are restrained by a narrowness of mind, unfortunately too common, and oppose to ardent, though generally thoughtful, enthusiasm, the question so powerful with many minds, the *cui bono* of the utilitarians. All that is not an

---

* We have thought it right to place after Bellot's Journal some notes full of interest, as well as several important letters, all relating to this voyage, and which we found in the same register. The author undoubtedly wrote these notes, and copied the letters, under the idea that they would serve as documents, and possibly find a place in the description of the voyage, which he intended himself to publish at a later date.

immediate profit, palpable, of a nature to be understood by the most vulgar mind—all, in short, which does not produce money—is branded as a wild dream—the project of visionaries—impossibilities! and still, as if bearing generous testimony, the result has nearly always belied their sinister predictions—the terrific presages which would have terrified men cast in an ignoble mould! Ferdinand and Isabella, engaged in the conquest of Granada, fearing a war with France, and casting a longing glance on the kingdom of Naples, were entreated by their councillors not to launch into dangerous and doubtful enterprises, leaving out of the question the impossibility suggested by the superstition and ignorance of the age: the treasures of the New World rewarded their perseverance.

These objections, for a discovery doubtlessly of slighter importance, were repeated in England, and the same band of economists cried aloud at the enormous expenses occasioned by the Arctic expeditions. The religious feeling, easily refuted by the Bible itself, furnished weapons to the adversaries. A few days later, without taking into account the treasures of geography and science laboriously amassed by the discoverers, a new and fertile field was opened up to commerce; and for more than a quarter of a century the whale-fisheries in Baffin's Bay and Lancaster Strait have given Great Britain several millions of money, and sailors hardened to the rudest toil. A reproduction at the present day of the easy answers which may be given to the arguments brought forward formerly, would be an insult to our generation —an attempt to gain converts. What nation in our day would place in the other scale these considerations, and allow them to outweigh the glorious privilege of discovering new countries? France, I repeat with pride, is not behindhand: to use a celebrated saying, she has always been rich enough to pay for her glory.

Christopher Columbus was refuted by the words of the Psalmist, that "the heavens stretch out over the earth like a curtain." A minister of the Gospel in England says that the heavens designedly raised up insurmountable barriers, and that persevering was tempting Providence; as if this struggle, this development of human faculties, were not the fairest offering laid on the altar—the burnt offering of the Bible.

The history of the attempts made by English navigators to discover the north-west passage, constitutes, it is true, if we take into consideration the final object proposed, a long succession of checks; but each of these checks is, in reality, a glorious struggle against the elements—against Nature herself; but, there must be no mistake on this point; this series of negative solutions must necessarily conduce to a definitive result.

It is equally glorious for geography to determine the limits of an empire, as to add new provinces to it; and in an age where so little is left us to do, it is a good deal to regulate the frontiers of the world!

A great noise is made about contests undertaken in the cause of liberty; as if the result of a struggle were not nearly always a change of characters, and the subjugation of one dominant race to another, which, in its turn, becomes oppressive: thus a long history of misery and mourning is perpetuated. True liberty can only be the result of an intellectual war, especially of the industrial war of nations. This war is carried on not with salvos of artillery, but by means of discoveries, geographical or otherwise. The sword will be converted into a ploughshare; powder will be only employed to tear from the earth treasures it conceals from us, and to destroy the ice of the two poles; and of all our permanent armies, the military marine will alone survive as instruments of peace, of civilization, of discovery, and as commercial police. *Si vis pacem, para bellum!* exclaims the wisdom of nations—a sentiment of distrust dictated by our antiquated policy, leagued with detestable traditions which alone are still able to render the axiom a truth. This is the way in which it ought to be interpreted: Be ready for war, but, in the sense of the spirit of your age, a commercial, industrial, and scientific, war. Commerce, ever commerce! our newspapers exclaim, with a disdainful air! Grocers! the baffled socialists howl. Yes! commerce, for the riches of a nation are represented by that which ensures its existence. I say commerce, because I am now acquainted with English manners, and to them I attribute the liberty which that great nation enjoys; for I am quite confident the more commercial a nation is, the more does it enjoy that liberty which offends none—that liberty where every man is free; in a word, liberty allied to *order*—a thing which we are inclined to consider a myth, like the philosopher's stone, and the quadrature of the circle. The day on which we become a little more commercial will open before us an era of happiness.

The phrase, "England is a nation of shopkeepers," is as insolent as it is unjust; for the mercantile idea bestowed on us the new world. Associated merchants were the first to discover the North of Europe, Novaia Zemlia, Northern Asia, the Arctic regions to the west of Groënland, Hudson's Bay, Vancouver's Land, and the Californias; it was a merchant, Mr. Booth, who fitted out Ross's expedition in 1829; merchants, Messrs. Enderby, sent vessels to the Antarctic Pole; an American merchant, Mr. Grinnel, equipped two vessels to search for Sir John Franklin; the Hudson's Company fitted out another; and the majority of those who subscribed for Back's voyages, and for our own, are merchants.

Yes! let us have the war that propagates, instead of the war that destroys! the war that gives bread to thousands before the Malthusian remedy, which relieves the world of its plethora! I only had time to take a hurried glance at the Crystal Palace, but I am now happy that our vessel and our first discovery bear the name of Prince Albert.

THE RELATIVE HAPPINESS OF THE ESQUIMAUX,
AND OF SAVAGES GENERALLY.

It is very true that happiness is relative; but this remark is far from clenching the pretended philosophic axiom, "The more a man restricts the number of his wants, the greater chance he has of satisfying them, and, consequently, of being happy." In the first place, the argument is bad; for, if we admit at the outset that happiness consists in the satisfaction of our wants, it would result, on the contrary, that the more wants we had to satisfy, the happier we should be: the sensualist, the gourmand, would rejoice greatly in having seven stomachs kindly allowing him the incessant tickling of his palate. This materialism would soon lead to the cessation of civilization. In the moral order, is it not evident, on the contrary, that the successive extension of wants not necessary to existence, but nevertheless imperious, has furthered the progress of the arts and sciences? Forget this consideration, and man would soon return to the level of the brute; for our sole wants would be reduced to food and clothing, which provident Nature has nearly universally placed within our reach. Do we not call this creative power Providence, because it provides for us? Are the Esquimaux more wretched than the Pechoras of Bougainville, in Terra del Fuego, who do not even employ their seal-skins or the feathers of aquatic birds to cover them?

With the Esquimaux, in their present condition, the satisfaction of their corporeal wants is the first condition; consequently, when I offered them beads, looking-glasses, and iron tools, the first, though exciting their covetousness in the highest degree, yielded to more useful objects. Dainties, which were offered to them, pleased them much; and they put a portion aside for their women, in spite of the inferiority which they regard them with: but this feeling did not go so far as to induce them to exchange these objects for hide or horn, whence they would derive a profit of greater advantage to them.

The idea of the character which the arts assume in civilization is now so thoroughly conceded, even by minds the least enlightened, that no one dreams, at the present day, of those senseless projects of suppressing luxury which flattered the envy of taproom agitators at the close of the last century. Modern reformers and economists, on the contrary, have justly considered that the greater the demand is, the greater will be the development of human activity.

The simplicity of Spartan manners is not adapted to our age; but it is said, "luxury is the rock on which liberty splits, and republics perished through corruption." This was true at a period when the continual wars of state against state, city against city, rendered it dangerous to foster that refinement produced in the habits of life by the cultivation of the arts; men attached to study and intellectual pleasures were reluctant to lay down their pen and quit their library for the sword and the battle-field: but at this period the ulcer of slavery infected the whole social body, and, as there was not a free scope, wants badly understood assumed a false direction; mechanical labour was left to the slaves as a degrading task, and the energies of those men who felt a necessity for activity were squandered in disgraceful inactivity, as they had no taste for literary pursuit or philosophic speculations. At the present day, on the contrary, what human aspirations would not be satisfied by mechanics, chemistry, or commerce? And, again, what is that ancient liberty whose praises are so universally sounded in our ears?—this pretended independence, so sensitive, so jealous of its privileges? What else is it than the realisation of the most perfect egotism?

The Esquimaux, for instance, who seem to

have no leaders, are independent; that is, there is no law obliging any one by his personal labour to contribute to the wants of the community in general, or of one of its members individually; but, on the other hand, there is no existing tie among them, and their dependence on their daily wants is the greater, because they are ignorant of that association of exertions which in organised society forms a solid body of the whole.

The legislators of whom we have just spoken were besides consistent to themselves and their laws, when they demanded the abolition of luxury, in order to maintain among the warriors a readiness to undergo the privations of a military life; but when they believed that they had suppressed corruption by suppressing those acts which effeminate, according to their idea, and destroy the taste for warfare, they took the cause for the effect. What enjoyment could be felt under this pretended liberty, gradually hardening the feelings, in the midst of that perpetual mistrust of the citizens toward each other, of that insupportable tyranny of the individual over the remainder of the community, and of the community over each? This species of liberty appears to me thoroughly opposed to Christianity — to the republic of Christ.

"Do not to others what thou wouldst not they should do unto thee," is a precept interpreted, "Let not that be done to another which it does not please thee to do." As for the divine recommendation, "Love thy neighbour as thyself," no one thought of it. Thus, beneath the shadow of the most detestable and hypocritical puritanism, individual liberty was paralysed, and they fancied they had produced universal liberty by the sum total of the despotism exercised on each individual.

As is the case in all savage nations, excepting the tribe which Cook saw in his third voyage, where the men were almost treated as slaves, a woman, among the Esquimaux, is the husband's first and most immediate servant. Their degradation, however, is not so great as among certain races; and certainly the treatment women experience from men is one of the features by the aid of which the ethnologist can classify the different peoples on the scale of civilization—not that I am of the opinion of those reformers who, if we carried out their system to the fullest extent, would give women in society the same functions and duties as men. Why, then, continually desire things contrary to nature? and why be so obstinate in comparing beings who are not comparable—fractions with different denominations? For my part, I consider that, as an exception to the rest of the animal race, woman alone is not exclusively the female of the male. She can move onwards, parallel, and side by side with man; and there is great common sense in calling either sex the moiety of the human race. The part which Providence has assigned to woman is as noble and essential as that of man, destined by nature for different labours.

It is, however, evident that wherever brute force prevails, the power of man, being exercised without moral control, will be felt in a manner more or less cruel in proportion as his mind is enlightened.

Nearly all the Esquimaux taken to England, after the first emotions of surprise, evinced a desire to return to their icy regions and their greasy huts. Nostalgia exists even among these poor creatures, made, like us, after the image of the universal Creator; and why should we feel astonishment at it? The nature of man, despite the pessimists, is essentially sympathetic; it is the divine stamp of his origin, and from the whole of our being rays are emitted which warm and animate the substance, and communicate to the most sterile rock a portion of our life—of our thoughts; and then we love this dead nature, as Pygmalion idolised his statue—the child of his creation. That which we love in it, which does not speak to the heart of the stranger, is a witness of the tears and joys of our infancy—is the confidant of our intimate thoughts; it is ourselves at every epoch of our past life. Are we not attached, after a voyage of a few months, to the most indifferent beings? The result of habit, you may say. Yes, but in this respect, what are all our sweet and durable emotions, if not pious habits? I cannot gaze without emotion on the stone benches—the young trees which I saw planted round our *Place d'Armes* at Rochelle. This stone is, in fact, a friend of long standing, who witnessed more than one of my boyish exploits, and was the accomplice of all my tricks. Thus, with the Esquimaux, the sight of the snow hut where he has passed the winter—the hole which protected him—is more pregnant with emotions than the pompous façades of our edifices. What do those rich palaces in which we have not lived say to us, or the lofty trees—the beautiful tufted chestnut trees— beneath whose shade our infancy has not sported? Ah! they can speak to our senses —to our mind—to our head: we admire them as the *chef d'œuvres* of man—as the witnesses of our own grandeur; but our heart finds them dumb. The savage as yet has no mind, and scarcely a head; but he has a heart, and it is by that he is really a member of the great human family.

The books of a school, about which I have had great illusions, and in which I still possess friends, hardly recognise as legitimate the imperfect and limited feelings which we have in our era of imperfect harmony. Our moral faculties are limited, less perhaps than, but in a manner equally real with, our physical faculties; the luminous rays must be concentrated on a point before developing any heat, and it is the same with our feelings. I do not believe in filial love, in paternal affection, in the love of the sexes, or in friendship: I do not believe, I say in their power, unless their forces converge on a very limited circle of objects. In despite of theories, our loving faculties cannot be elongated or extended infinitely without a risk of breaking; and I doubt whether man ever arrives at that state of cosmopolitism which would cause him to love all his fellows equally. Would a change in our social organisation produce a change in our moral organisation? Who can say so? or, rather, who can demonstrate it? I believe that family, friends, city, country, only exercise a great empire over us because they are within our reach; and in infinity we can only serve God with devotion, because he is *one*. Do not fancy that by this I recommend exclusivism: no—a thousand times, no; but let us examine if the theory is not palpable by its actual results—if it is not displayed in our every-day life. Do we not daily hear it said of a man who loves all the world, "He has a heart like an hospital, open to the first comer?" and who cares for the friendship of such a man? not through jealousy because we wish to be the sole object of affection, but because we feel that his friendship must dissolve in the midst of this mass of affections, like a grain of vermilion in a glass of water. Do we not read every day in the papers that such or such a state, which is incessantly acquiring new possessions, will eventually be dismembered owing to this very aggrandisement? Can anything which is too great or phenomenal live? No; it is not in nature. It is better to lop a tree in the spring, if you wish it to put forth vigorously. Cherish in the innermost sanctuary of the heart our nationality, and thus we shall be able, without fear or reproach, to sacrifice on the altar of charity this blind and misshapen nationalism.

The instructions given to Sir J. Franklin,

as well as to the officers sent in his research, order them to avoid most carefully any acts of hostility, even against a nation with which the Government might have declared war during their absence. I cannot resist a feeling of pride in drawing attention to the fact, that France was the first to give this noble example of the respect due to science, by not rendering discovery ships amenable to the usual laws of warfare.

When, on the death of Cook and Clarke, the two ships, the *Resolution* and *Discovery*, returned to England in 1780, Captain Gore, who had assumed the command of the expedition, was fearful lest the fruits of this long and laborious voyage (1776—1780) might not escape the chances of a contest; but the instructions of a king who protected geography and navigation ordered his officers to keep the peace, in the event of their meeting with Cook's vessels. These instructions were found on board some of our vessels which were captured by the English.

LLTTER TO M. BARROW, ESQ.,
AT THE ADMIRALTY.

ABERDEEN, *May*, 1851.

MY DEAR SIR,

I do not like to quit England without an attempt to express my gratitude for your kind behaviour in my affairs. I have already received from all the members of the Admiralty marks of attention and kindness, which fill me with confusion. I can only ascribe to my uniform the honours which have been paid me. During my entire stay at Aberdeen, Lady Franklin and Miss C—— have continually honoured me with delicate attentions, of which I am proud and happy, although they are scarcely deserved. What have I done but what was quite natural in asking to share in the honour of the dangers to which so many brave men are about to expose themselves?

I wrote to M. le Ministre de Marine, explaining to him that an officer could not possibly have refrained from such an expedition through fear of a few privations. After having been witness of that noble grief—of that indefatigable perseverance, which allow Lady Franklin no rest, no repose, I feel that I would have gone on with my eyes closed. Such incessant devotion would inspire the deepest sympathy in any man. Who would not feel himself filled with a holy ardour at the sight of the labours and mighty exertions of this devoted wife? I am glad of the delay we have experienced here, because it has allowed me to appreciate the noble feelings of these ladies; and I have been enabled to realise more intimately the ardour which they bring to bear on the execution of their projects. I now fancy I form a part of the family.

You have been so kind to me already, that I am almost ashamed to ask fresh services at your hands. I shall be doubtlessly deprived of opportunities to write to my family, and I would ask you, whenever an English paper gives any news about the *Prince Albert*, to cut out the paragraph and send it under cover to Madame Bellot, Rochefort-sur-Mer. I forget who it was that said four-fifths of a letter always serve as preface to the writer who only mentions the real object of his epistle at the end. Believe in the truth of this, my dear Sir, and allow me here to express the feelings of sympathy and affection with which you inspire me. French petulance, perhaps, causes me to leap too soon over the usages of *convenance;* but the kindness of Captain Hamilton and yourself have affected me in a way I cannot describe: were it not for the difference in our ages, and my humble epaulette, I would beg you to tell Mr. Hamilton how I feel affected toward him. Be kind enough to present my humble respects to all the members of the Admiralty.

Will you allow me to write to you now and then? I take the liberty of sending you my portrait, that you may at times call to mind a person who will never forget your good offices.

     I am, dear Sir, &c.

LETTER TO M. XAVIER MARMIER,

KEEPER OF THE ST. GENEVIEVE LIBRARY.

ABERDEEN, *May*, 1851.

MY DEAR SIR,

Once again I am setting out on one of those long voyages which you taught me to love—that bird-of-passage nature, which forms a bond of sympathy between us, carries me off once more. You will understand the object that summons me, and your vows, I doubt not, will follow the seaman who is guided by a feeling of enthusiasm and ardour alone.

You know, my dear Sir, that Sir J. Franklin set out in 1845, in search of that north-west passage of which illustrious English captains often fancied they held the keys; the dauntless and skilful labours already accomplished, the experience acquired, a perfect acquaintance with the regions which the *Erebus* and *Terror* must explore, suffered little doubt to be entertained but that Sir John would meet with success, if the passage really existed. The reflecting boldness of this courageous captain has, however, induced him to risk fresh dangers, as no news of this expedition has been received since his departure. The scientific world was startled by the disappearance of the two vessels; and the English Government, more interested than any other in learning what had become of them, sent successive expeditions in search of these unfortunate men. The worthy helpmate of Sir J. Franklin—the devoted wife, whose thoughts had constantly followed him in his labours, as well as his dangers—cannot remain inactive in the midst of the preparations making in England: her ardent affection causes her to regard as insufficient the conscientious measures taken by the Government, and the greater portion of her fortune is the reward she offers to speculators—to the less disinterested devotion of those who could follow the traces of Sir J. Franklin and his companions. The other portion is devoted to the equipment of an auxiliary expedition, which the sublime impatience of Lady Franklin sends out under the command of an officer of the English navy, who has voluntarily offered his services. While ships, after doubling Cape Horn and passing through Behring's Straits, are seeking the absent on the shores of the Arctic Ocean, Captain Austin, with two vessels, has been sent to examine the spots where the *Erebus* and Captain Franklin were to stop. The American Government had hinted a design to send some of its vessels : it hesitates at the moment of accomplishing its promise: a simple citizen, Mr. Grinnel, will not suffer the engagement of his nation to be broken, and at his own expense sends out an auxiliary expedition: and, lastly, a ship commanded by Penny, a captain of a whaler, also starts in 1850. Through accidental circumstances, the *Prince Albert* was compelled to put back without obtaining any satisfactory results, after four months' absence. A few pieces of rope, provisions—evident proofs of Franklin having passed Cape Riley—cause the most lively emotion in the English public. The clumsy drawings traced by the Esquimaux prove that some, if not all, allowing the worst, had escaped from the disaster of the *Erebus* and *Terror*. Perchance, at the present moment, the bold sailors who left during the last year have attained the object of their labours. But, in such a struggle, can there be any jealously, any rivalry, save that of the exertions they desire to make? What matters the name of the conqueror, provided the victory is gained?

The *Prince Albert*, then, will renew this year the exertions which she could not carry out during the last. We were at Monte Video when the various expeditions were made known to us, and I cursed my absence from the point of departure. Thus, as you may imagine, I leaped with joy, on seeing that Lady Franklin had not lost courage, and was equipping the *Prince Albert* once again.

Our Minister of the Navy understood that men like Sir John Franklin are not merely citizens of a nation, but of the whole civilized world; and that, as all nations had a share in the success and glory of these illustrious men, all ought to be represented in such an expedition as this. Lady Franklin has received propositions on all sides. Last year two captains requested to form part of the expedition under the orders of the captain of a whaler; and this year Commander N. offered to furnish £500 if the command of this schooner were given him. The engagements Lady Franklin had already entered into with Captain Kennedy did not allow her to accept this generous offer. The evidences of sympathy on the part of the French navy inspire the English Admiralty with the greatest gratitude, and I am compelled, willy nilly, to accept their thanks—I, the modest and humble representative of our uniform. Several newspapers, and especially the *Morning Herald*, see, in our participation in these glorious efforts, one of those acts tending to cement among all nations that union which is the object, and of which the Universal Exhibition is one of the means. This is giving too great a range to individual and limited demonstrations; and, without touching on thoughts of so elevated a nature, who would not feel his strength redoubled at the sight of the courage of Lady Franklin and a devoted niece, who has for a long time shared in these labours. Assisted by a few friends, Lady Franklin and Miss C—— keep up a correspondence extending to all parts of England—I may say, of Europe; for, in addition to the information they require, and which they ask from all those to whom an Arctic voyage has given some experience, these ladies receive all sorts of suggestions—some excited by real interest, which strives to find an outlet; others by a necessity, more or less disinterested, of an individual trial. One day it is an artificial method to keep up heat by means of smoke, and the light of a lamp hidden under the clothes; on another, information obtained by mesmerism as to the spot where the absent men are at this moment. I intend to give you, in a letter to follow this shortly, some details about the expeditions already made. Owing to some very unlucky circumstances, no one has yet been able to visit the spot where it is most probable that some information at least may be obtained about the crews of the *Erebus* and *Terror*. After entering Davis's Straits and Baffin's Bay, pass Lancaster and Barrow Straits, and penetrate into Prince Regent Gulf; you will then find, on the eastern coast, a spot marked Fury Beach. It was here that Captain Parry lost, in 1825, one of the vessels in which he made his third voyage into the Frozen Sea. Captain Franklin, who knew he would find there immense stores, must assuredly have gone there, admitting that he found himself under the necessity of abandoning his ships. His instructions told him to try and pass the south-west of Cape Walker: other routes were pointed out in the case of an impossibility; but those who know Captain Franklin are certain he would follow his instructions to the letter: but, as the winter of 1845 was extraordinarily rigorous, it is evident that he would think about abandoning his vessels at the moment when provisions began to fail him; and, as he was perfectly acquainted with the country, he would try to make his way to Prince Regent's Straits— the nearest to Europe; and, consequently, he would most naturally expect succour there.

Now, this year it becomes the more essential to visit these very points, for the vessels which have been there since last year are ignorant of the return of the *Prince Albert*, and will not dream about visiting those places which she was sent out to explore. This expedition will touch at Griffith Island, where Captain Austin will have left some notice of his own movements. In the event of our predecessors having spared us the trouble, the *Prince Albert* will go as far south as possible in Prince Regent's Gulf, either to Brentford or Cresswell, and thence boats will be sent along the channel formed by the sea along the coast, and which will allow them to go a great distance.

As it is possible that want of success or the ice may keep us out till 1852, the schooner is provisioned for two years. The system proposed in previous years has not been adopted, consisting in throwing out from a balloon thousands of bills announcing the efforts we are making to reach the captives. These balloons would also allow us to take a hurried glance over a large tract of country. Without discussing the merits of this idea, it was thought that the small number of our crew, and the size of our vessel, would prevent a complication of experiments which generated and annihilated each other reciprocally. A fancier in England has provided me with six carrier pigeons, which we shall let at liberty either at points arranged beforehand, or at places where we shall have a chance of announcing any important facts.

The results already obtained allow us to calculate, at least till the contrary is proved, on the effective reality of this mode of transmission. One of these birds has already stood the test, for there are good reasons for believing that it is one of those liberated last year by Captain J. Clark Ross. Perhaps, in any case, we shall thus have opportunity of solving a curious problem.

For our voyages of exploration, we are provided with a magnificent mahogany whaleboat, and a gutta-percha boat. Captain Kennedy has also a canoe, twenty two feet long and two wide, made of block tin, and covered with a layer of gutta percha, after the model of the Esquimaux boats. This canoe, which is entirely covered, has only a circular hole in the centre: a single person manages it by means of a double paddle. The two ends, being rounded, give it an impulsive power. To the edges of the hole is attached a leather, in the shape of a *blouse*, which the sailor puts on, and buttons up to the chin, which renders any immersion impossible. The canoe, thus managed, floats in a few inches of water; and, after being used to clear the space between two lumps of ice, will carry lines of communication from one point to another: it can also be used as a tent during the night. Lieut. Halkett offered me, for my own use, a waterproof cloak, which can be converted into a canoe by filling it with air between the lining and the exterior. These two objects are of immense value in a country where the want of wood renders it impossible to form any species of raft. Sir J. Franklin almost perished of hunger in 1819, on the banks of the Coppermine River, owing to the want of a simple plank, which would have enabled him to join a friendly tribe a few miles off. Everybody, indeed, seems anxious to satisfy our wants, or even our pleasures; for Prince Albert has made a present to the expedition of a magnificent organ, which in itself forms an entire orchestra: the Admiralty has provided us with a great quantity of pemmican. You are aware that this species of food contains, in a small bulk, a great quantity of nutritious matter, and hence becomes very useful, owing to its facility of transport.

We should really not have known where to stow all that kindness presses upon us, were we not vowed to the teetotal principle; that is to say, the most absolute temperance. No wine, no beer, no cider in our stores: no spirits, save in the apothecary's shop. (O, Bacchus! O, Pomona! turn away your eyes, ye vengeful divinities of my dear Saintonge! Spare your faithless child!)

This precaution, founded on the observations of persons most experienced in the Arctic regions, will doubtlessly impart fresh lustre to our expedition. Hardy Scots of the Orcades, or Shetland Isles, who formed part of the previous expeditions of Rae, Richardson, and Franklin, or tried by numerous voyages in search of whales, form a chosen crew. Mr. John Hepburn, who followed Franklin in his examination of the Coppermine and Mackenzie Rivers, has arrived in all haste from Van Diemen's Land, to furnish a fresh proof of his devotion to his old captain. Mr. Leask, pilot of the North Star, who knows the Baffin and

Barrow Straits as well as you do your library, is our ice-master. At our head is Captain Kennedy, a captain in the Hudson's Company's service; a man of an ancient stock; a scion of those Puritans whose dauntless courage has its source in the most lively faith; one of those models from whom Cooper has taken his Pathfinder. Alone, in the midst of these men, tried by incredible sufferings, I bring, instead of experience, a boundless ardour; but I have confidence. Have we not the justice of our cause to back us up?

Time fails me, my dear Sir, to give you new details, both about the real chances of our expedition, and about the methods chosen to ensure its success. A few whalers will doubtlessly afford us opportunities to write to France, at least up to the month of August. I will keep you *au courant* of our progress: a few lines, traced in haste—a little gossip about all sorts of things—are all I can offer you in exchange for your pleasant conversation about South America: those long discussions, over which you scattered with a lavish hand the flowers of your poesy; but friendship is indulgent, and you will not disdain, in return for your rich garlands, the dry and indigestible biscuit of the sailor.

Believe me, &c.

TO THE MINISTER OF THE NAVY.

ABERDEEN, *May* 15, 1851.

SIR,

Thanks to the kindness with which you met my application, and the *empressement* with which the officers of your staff arranged my affairs, I was enabled, four days after the receipt of Lady Franklin's letters at Rochefort, to present myself to that lady and the Admiralty in London. The honest and faithful explanation of the privations to which we are exposed on board the *Prince Albert*, a brig of 90 tons, they expected would have disheartened me, and no one anticipated that I would come; but would it have been possible for a French officer to draw back on account of a few dangers to be incurred? and, once having obtained your approbation, it was not to be thought that I would hesitate. The aspect of that noble grief, of that indefatigable devotion, which suffered Lady Franklin to take no rest, would arouse courage and enthusiasm even in men hitherto ignorant of such feelings.

I told both Lady Franklin and the members of the Admiralty how you intended to evince your sympathy for the English navy and the bold sailors whose loss we all regret, by allowing one of your officers to take part in this expedition.

Admiral Sir F. Beaufort, head of the hydrographical department; Captain Hamilton, permanent secretary of the Admiralty; Mr. Trotter; Mr. Barrow, son of the illustrious savant of the same name, have not ceased to overpower me with kindness and delicate attentions, which I naturally referred to my uniform and the sympathies I represented.

The First Lord of the Admiralty, Sir E. Baring, and Admiral Dundas, requested that I should be presented to them, in order to thank the French navy, in the person of one of its most humble members, for the exertions it is about to share in.

Lady Franklin wished to pay the expenses of my stay in London and Aberdeen, as well as my outfit; but, although my travelling expenses were insufficient, I immediately explained the entirely disinterested position I intended to assume in this expedition—the only position which could satisfy your intentions, and maintain the dignity of my character on board the *Prince Albert*.

As I am unaware what has been hitherto published in France about the researches already made, I add to my letter a hurried sketch of the preceding voyages. I cannot refrain either from the pleasure of sending you extracts from two English papers, in which our participation is sokindly appppreciated.

Will you permit me, Sir, to ask you to prolong the payment to my family of a portion of my pay, as I requested on the 8th of May? I think I may calculate on this favour; and I beg you to accept the assurance of my most profound respect,

J. B.

22nd May.—We have just started for the Shetland Islands, and I reopen my letter to inform you, Sir, of two fresh acts of kindness paid to our navy. Mr. Arthur Thompson, vice-consul of France, and one of the richest bankers in Aberdeen, knowing that I left to-day, has hoisted flags on his house. The warm expressions of sympathy which Mr. Thompson has always uttered, his affection for our naval service, and this fresh act of politeness, compelled me to go in person and thank him.

By Lady Franklin's orders, the French flag was hoisted simultaneously with the English on board the *Prince Albert*, as a sign of our heartfelt union; and the people of Aberdeen ran down to the quays and saluted it with shouts.

Lastly, the English Admiralty has sent letters to the captains of all the ships of war in the northern seas, recommending us specially to their notice.

LETTER WRITTEN BY BELLOT TO HIS FAMILY.*

Wednesday, 10th September, 1853.

MY DEAR AND EXCELLENT FRIENDS,

If you receive this letter I shall have ceased to exist, but shall have quitted life in the performance of a mission of peril and honour. You will see in my journal, which you will find among my effects, that our captain and four men were necessarily left behind in the ice to save the rest; so, after effecting that, we were compelled to go to the assistance of these worthy fellows. Possibly I had no right to run such a risk, knowing how necessary I am to you in every way; but death may probably draw upon the different members of my family the

* This touching farewell letter, written by Bellot before one of the most perilous events of his voyage (see Journal, date 10th September), and which would only be delivered if the heart that inspired it had ceased to beat, was not recovered by his family till after the loss of the heroic young man, two years after its date.

consideration of men, and the blessings of Heaven.

Farewell! to meet again above, if not below. Have faith and courage.

God bless you!

J. BELLOT.

THE BELLOT TESTIMONIAL FUND.

As soon as the melancholy news reached England that the dauntless Frenchman who had so nobly undergone privation, fatigue, and danger, in the cause of humanity and science, had sacrificed his life in the performance of his arduous duties, and when the first feelings of heartfelt sorrow had become merged in admiration of his heroism, it was felt that it behoved us to recognise his merits in the manner which, had the choice been left to himself, he would doubtlessly have found most consonant with his feelings. Strong affection for his parents and his family is evinced through the whole of Bellot's journal and correspondence, and their welfare was his constant care. By his lamented death, they would be deprived of many of the comforts of life, and their feelings of sorrow would be augmented by the reflection that their mainstay had been taken from them at a moment when they could least afford to lose it. Englishmen acted worthily of themselves: a meeting was immediately convened at Willis's Rooms (Nov. 4th, 1853), Sir Roderick I. Murchison in the chair, when the following resolutions were unanimously agreed upon:—

1st.—Moved by the Right Hon. Sir JAMES GRAHAM, Bart., M.P., First Lord of the Admiralty, F.R.S., F.R.G.S., &c.;

Seconded by Rear-Admiral Sir W. EDWARD PARRY, D.C.L., F.R.S., F.R.G.S.—

"That this meeting, composed of various classes of Englishmen, being anxious to mark their deep sense of the noble conduct of Lieutenant Bellot, of the French Imperial Navy, who was unhappily lost in the last Arctic expedition in search of Sir J. Franklin, resolve that their countrymen be invited to unite with them in promoting a general subscription for the purpose of erecting a monument to the memory of that gallant officer, to be placed at an appropriate spot at or near the Royal Hospital of Greenwich."

2nd.—Moved by Colonel SABINE, R.A., F.R.S., F.R.G.S.;

Seconded by Capt. R. FITZ-ROY, R.N., F.R.S., F.R.G.S.—

"That the surplus of the subscription, after defraying

the cost of the monument, be invested for the benefit of the members of the family of Lieutenant Bellot."

3rd.—Moved by JOHN BARROW, Esq., F.R.S., F.R.G.S.;

Seconded by Capt. E.A. INGLEFIELD, R.N., F.R.S., F.R.G.S.—

" That it be an instruction to the sub-committee to communicate with the municipal authorities of other ports of the United Kingdom, and with the Naval Commanders-in-Chief, in order to make the subscription general, and particularly among the seafaring population."

4th.—Moved by Capt. W. A. B. HAMILTON, R.N., Secretary to the Admiralty, F.R.G.S.;

Seconded by Capt. E. OMMANNEY, R.N., F.R.G.S.—

" That the following gentlemen be a sub-committee to carry out the objects of this meeting :—Sir Roderick I. Murchison (chairman), Earl of Aberdeen, Earl of Ellesmere, Sir James Graham, Sir R. H. Inglis, Admiral Beaufort, Colonel Sabine, Ch. Wentworth Dilke, John Barrow, and H. Robertson, Esqrs.; Captains Fitz-Roy, W. H. Hall, Ommanney, and Inglefield, R.N., with the Rev. G. C. Nicolay and Dr. Norton Shaw as secretaries."

5th.—Moved by Capt. HORATIO T. AUSTIN, R.N., C.B.;

THE BELLOT TESTIMONIAL FUND. 385

Seconded by Lieut.-Col. NEIL CAMPBELL, F.R.G.S.—

"That the thanks of this meeting be presented to the Public Press for its voluntary support of the Bellot testimonial, and for its free and liberal insertion of advertisements."

6th.—Moved by Sir J. GRAHAM, Bart., &c.; Seconded by Sir R. H. INGLIS, Bart., M.P., &c.—

"That the cordial thanks of the meeting be voted to the Chairman."

By the extreme kindness of Sir R. I. Murchison,[*] to whom the editor applied for some information on the Bellot testimonial, he is enabled to announce the complete success of the appeal to the English nation on behalf of Bellot's family. The sum obtained, slightly exceeding £2000, enabled the committee to divide a large portion of the subscriptions among the five sisters of the deceased, each of whom will eventually

[*] The following extract speaks volumes for the extreme interest Sir R. Murchison felt in the subject:—"I willingly afford you any information respecting the testimonial raised, and the monument about to be erected, to the memory of my valued young friend, Lieut. Bellot."

realise about £300 — the two brothers having been provided for by the French Government. The remaining sum will be applied to the erection of a durable monument of granite, on which the name of Bellot will be engraved in large letters, to commemorate the admiration felt by Englishmen towards the lamented French officer.

The committee having entrusted to Sir R. Murchison the task of carrying out their wishes and those of the subscribers Sir Roderick informs us that the monument is to be an obelisk of Aberdeen granite, of the height of 34 feet, which will be placed in such a position on the bank of the Thames at Greenwich, that every one passing along the river or the quay must see it. It was no easy matter to secure such a site, for no English officer—not even Nelson—has had a similar mark of respect paid him on the *exterior* of Greenwich Hospital; but Sir Roderick was indefatigable, and being supported by the late First Lord of the Admiralty, Sir James Graham, he eventually succeeded in obtaining the requisite permission from the commissioners of the Royal Hospital, whose architect, Mr. P. Hardwick, R.A., kindly designed the form of the monument.

An unexpected delay on the part of the granite worker at Aberdeen, who contracted to execute the task, has alone prevented the obelisk from being already erected; but Sir R. Murchison confidently trusts that everything will be completed this ensuing autumn. The obelisk will stand insulated on the little wharf of the hospital, and in front of one of the central gates; so that, when viewed from the Thames, it will have in the background the fountain of the western esplanade or green, and the observatory on the hill.

Sir R. Murchison concludes his interesting communication with the following sentence, which proves that Bellot must indeed have been no ordinary man:—" It is hoped that this arrangement will prove satisfactory to the subscribers, and will testify to the French nation the sincere desire of Englishmen to do honour to the memory of the *good and intrepid Bellot.*"

We have appended a list of subscribers to the Bellot testimonial, for we think that their liberality deserves a permanent record; and we have to express our thanks to Dr. Norton Shaw, R.G.S., for enabling us to make it up to the latest date.

NAMES OF SUBSCRIBERS.

(Received at the Apartments of the Royal Geographical Society, 3, Waterloo Place, up to Nov. 21st.)

	£	s.	d.
Sir Roderick Impey Murchison, D.C.L., M.A., F.R.S., Vice-President R.G.S., *Chairman*	25	0	0
Aberdeen, the Earl of, K.T., F.R.S., F.R.G.S.	25	0	0
Acland, Sir Thos. Dyke, Bart., M.P., F.R.S., F.R.G.S.	5	0	0
Addington, H. U., Esq.	5	0	0
Aitchison, Capt.	3	3	0
Alston, Mr. A. H. (*Phœnix*)	2	0	0
Angerstein, Col. J. J. W.	5	0	0
Anonymous, Llanelly, per *Times* newspaper	1	0	0
Archer, H.M.S., Commander Strange, and the Gun-room Officers. One day's pay	3	16	6
Arrowsmith, John, Esq., F.R.G.S.	3	0	0
Austin, Capt. Horatio T., C.B.	2	2	0
Bacon, Thomas, Esq.	1	0	0
Baillie, Alex., Esq.	5	0	0
Baillie, David, Esq., F.R S., F.R.G.S.	5	0	0
Baker, Capt. Thomas, H.C.S.	1	1	0
Barnett, Capt. Edward, R.N., F.R.G.S.	1	0	0
Barrow, Sir George, Bart.	5	0	0
Barrow, Henry, Esq.	1	1	0
Barrow, John, Esq., F.R.S., F.R.G.S.	20	0	0
Beaufort, Rear-Admiral Sir F., K.C.B., F.R.S., F.R.G.S.	10	0	0
Becher, Capt. A. B., R.N., F.R.G.S.	1	1	0
Beechey, Capt. W. F., R.N., F.R.S., F.R.G.S.	1	1	0
Bell, J. D., Esq.	1	1	0
Best, Capt. the Hon. Thos.	1	1	0
Biggs, J., Esq.	2	2	0
Bird, Wm. Wilberforce, Esq., F.R.G.S.	5	0	0
Blair, Lieut.-Col. W. H. S.	5	0	0
Borland, Dr., Inspector-General of Army Hospitals	1	1	0
Bourdin, Mr.	1	1	0
Bowles, Vice-Admiral Wm., C.B., F.R.G.S.	5	0	0
Boyd, Mark, Esq.	2	2	0

NAMES OF SUBSCRIBERS.

	£	s.	d.
Bracebridge, C. R., Esq., F.R.G.S.	1	1	0
Breadalbane, Marquis of, K.T., F.R.S., F.R.G.S.	20	0	0
Brent, Geo. S., Esq., Deputy Coroner for Middlesex, F.R.G.S.	0	5	0
Brisbane, General Sir Thos. M., Bart., G.C.B., F.R.G.S.	5	0	0
Brock, Capt. Thos., R.N.	0	10	0
Brockedon, Wm., Esq., F.R.S., F.R.G.S.	5	0	0
Brooking, T. H., Esq., F.R.G.S.	5	0	0
Brown, John, Esq., F.R.G.S.	5	5	0
Brown, Robert, Esq., D.C.L., F.R.S., F.R.G.S.	5	0	0
Browne, Major-General Sir Henry, K.C.H.	2	0	0
Buckingham, J. Silk, Esq., F.R.G.S.	1	1	0
Buckley, Lady Catherine	2	0	0
Calder, Alexander, Esq.	2	2	0
Campbell, Capt. F. A., R.N.	1	1	0
Campbell, Lieut.-Colonel N., F.R.G.S.	2	2	0
Campbell, Major-General P., R.A.	2	0	0
Carr, Comm. Washington, R.N., F.R.G.S.	1	1	0
Cart, A., Esq., of Paris	1	0	0
C. C., per Hoare and Co.	5	0	0
Chamberlain, Comm.	1	1	0
Charlewood, Capt.	1	1	0
Chermside, Capt., R.A.	1	1	0
Clarendon, the Earl of, K.G., F.R.G.S.	25	0	0
Clerk, W., Esq.	1	1	0
Clive, Hon. Robert, M.P., F.R.G.S.	20	0	0
Coffin, Capt. Crawford, R.N.	1	1	0
Colchester, the Right Hon. Lord, Capt. R.N., D.C.L., Vice-President R.G.S.	5	0	0
Collier, C., Esq., M.D., F.R.S.	2	2	0
Combe, Harvey, Esq.	5	0	0
Commerill, Lieut., R.N.	10	0	0
Cooke, Comm. J., R.N.	1	1	0
Coote, C. Chidley, Esq., F.R.G.S.	1	1	0
Copley, Sir J. W., Bart., F.R.G.S.	25	0	0
Cotton, Capt. H. E., R.N.	2	0	0
Cracroft, Miss Sophia	2	0	0
Cracroft, Capt., R.N., F.R.G.S.	1	0	0
Cracroft, Mrs. Peter	2	0	0
Craigie, Capt. Robert	1	1	0
Cresswell, Lieut., R.N.	2	0	0
Cumming, R., Esq.	0	5	0
Davis, Sir J. F., Bart, F.R.S., F.R.G.S.	5	0	0
Delafield, W., Esq.	5	0	0
Denman, Capt. the Hon. J., R.N., F.R.G.S.	2	0	0
Devonshire, Duke of	25	0	0
Dickinson, F. H., Esq., F.R.G.S.	1	0	0
Dilke, C. Wentworth, Esq., F.R.G.S.	5	5	0
Dilke, C. W., Esq., F.R.G.S.	3	3	0
Dobinson, Wm., Esq.	1	1	0
Douglas, J. Stoddart, Esq., Lieut. R.N.	10	10	0

NAMES OF SUBSCRIBERS.

	£	s.	d.
Douglas, Col. Monteith, C.B.	20	0	0
Dover, J. W., Esq., F.R.G.S.	5	0	0
Downe, the Viscount, F.R.G.S.	10	0	0
Drake, Capt. John, R.N.	1	1	0
Drummond, Mr. (*Phœnix*)	0	10	0
Duff, Gordon, and Co.	2	2	0
Duncombe, Hon. Ar., R.N.	5	0	0
Dundas, the Right Hon. Sir David, F.R.G.S.	5	0	0
Durkin, H., Esq.	1	1	0
Edgell, Capt., R.N.	1	1	0
Edwards, Capt. H.	2	2	0
Edwards, Henry, Esq., F.R.G.S.	1	1	0
Eliott, Capt. George Angs., R.N.	1	0	0
Ellesmere, the Earl of, D.C L., President R.G.S.	25	0	0
Elliott, Lieut., R.N. (*Phœnix*)	2	0	0
Evans, D. M., Esq.	1	1	0
Evans, Capt. G., R.N., F.R.G.S.	1	0	0
Farley and Boyes, Messrs.	2	2	0
Farley, Wm., Esq., Master, R.N., Newton Abbott	1	0	0
Faraday, M., Esq., F.R.S.	2	0	0
Farrer, Jas. Wm., Esq.	2	0	0
Fawckner, Mr. (*Phœnix*)	0	10	0
Fayrer, Comm. Robert, R.N., F.R.G.S.	1	0	0
Ferguson, R., Esq.	1	0	0
Fitz-Roy, Capt. R., R.N., F.R.S., F.R.G.S.	2	0	0
Forster, John, Esq., M.P.	5	5	0
Fowler, R. N., Esq., M.A., F.R.G.S.	1	1	0
Franklin, Lady	25	0	0
Gage, Admiral Sir Wm. H., G.C.H., K.C.B., F.R.G.S.	5	0	0
Gassiott, J. P., Esq., F.R.S.	5	5	0
Gell, Mrs.	1	1	0
Gloucester and Bristol, Bishop of	5	0	0
Glyn, G. C., Esq., M.P.	10	0	0
Gordon, Capt. G. T , R.N.	1	1	0
Gordon, H. G., Esq.	2	2	0
Gore, Rev. George	1	0	0
Gowen, J. R., Esq., F.R.G.S.	2	2	0
Graham, Rt. Hon. Sir James, Bart., M.P., F.R.S., F.R.G.S.	25	0	0
Graham, Prof., F.R.S.	3	3	0
Grattan, T. C., Esq.	1	1	0
Griffith, C. Darby	10	0	0
Grosso, Mr. R.	2	0	0
Gurney, Miss Anna, of N. Reps	5	0	0
Haden, F. S., Esq.	2	2	0
Halford, Rev. Thomas, M.A., F.R.G.S	5	0	0
Halkett, P. A., Esq., R.N., F.R.G.S.	20	0	0

NAMES OF SUBSCRIBERS.

	£	s.	d.
Hall, Capt. W. H., R.N., F.R.S., F.R.G.S.	1	1	0
Hamilton, Capt. W. A. B., R.N., F.R.G.S.	5	0	0
Hancock, E. T., Esq	2	2	0
Hand, Capt. G., R.N., F.R.G.S.	1	1	0
Harker, Rev. G.	1	1	0
Harposel, F. R., Esq.	1	1	0
Harris, Mr. (*Phœnix*)	0	10	0
Harris, G. F., Esq., F.R.G.S.	1	1	0
Hawley, Mr. (*Phœnix*)	2	0	0
Hay, John H., Esq.	3	3	0
Herbert, Rear-Admiral Sir Thos., M.P., K.C.B., F.R.G.S.	5	0	0
Hicks, J., Esq., M.R.I.	1	1	0
Higgins, M. J., Esq.	5	0	0
Hills, Mr. (*Phœnix*)	1	0	0
Hobbs, J. S., Esq., F.R.G.S.	1	1	0
Hokin, Comm. C. L.	1	0	0
Holland, Sir Henry, Bart., M.D., F.R.S., F.R.G.S.	5	5	0
Holman, Dr. (*Phœnix*)	2	0	0
Hopkinson, W., Esq.	0	10	0
Ditto, for "the Family"	0	10	0
Horner, Rev. John M.	1	0	0
Howard, Capt. the Hon. Edward	1	1	0
Howden, Right Hon. Lord	10	0	0
Hudson, Capt. J., R.N.	1	1	0
Hughes Hughes, W., Esq.	1	1	0
Hurry, Edward, Esq.	1	1	0
Hutchinson, Col. H.	1	0	0
Hutchinson, Hon. R. H.	1	0	0
Ick, Rev. W. R., Syglesthorpe	1	0	0
Inglefield, Capt. E. A., R.N., F.R.S., F.R.G.S.	5	0	0
Inglis, Sir R. H., Bart., M.P., F.R.S., F.R.G.S.	5	0	0
Irving, Dr. E. G., R.N., F.R.G.S.	0	5	0
Irving, Thos., Esq., F.R.G.S.	0	5	0
Jesse, Captain	1	1	0
Johnstone, Colonel Geo.	1	1	0
Kaler, E., Esq.	1	1	0
Kennedy, William, Esq., of the *Isabel*	5	0	0
Knudtzen, Gorgau, Esq.	5	0	0
Laird, M'Gregor, Esq., F.R.G.S.	5	0	0
Lansdowne, the Marquis of	25	0	0
Lee, Thomas, Esq., F.R.G.S.	1	0	0
Leigh, Right Hon. T. P.	10	0	0
Levesque, Peter, Esq., F.R.G.S.	10	0	0
Lindsay, H. H., Esq.	1	1	0
Lloyd, —, Esq.	1	1	0
Lloyd, T. Davies, Esq.	1	1	0

NAMES OF SUBSCRIBERS.

	£	s.	d.
Londesborough, Lord, F.R.S., F.R.G.S.	25	0	0
Luce, J. T., Esq.	1	0	0
M'Cormick, Dr. (*Phœnix*)	2	0	0
M'Donnell, Mr. (*Phœnix*)	1	0	0
Maitland, Vice-Admiral the Hon. Sir Anthony	5	0	0
Majendie, Ashhurst, Esq., F.G.S.	5	0	0
Majendie, Mrs. Ashhurst	1	0	0
Malmesbury, the Earl of	5	0	0
Mangles, Capt. J., R.N., F.R S., F.R.G.S.	5	0	0
Manson, Mr. Donald (*Phœnix*)	0	5	0
Mare, C. J., Esq.	50	0	0
Martin, Rev. J. W., F.R.G.S.	1	1	0
Merivale, Herman, Esq., F.R.G.S.	2	0	0
Mills, E. W., Esq.	10	0	0
Milne, Alexander, Esq., F.R.G.S.	2	2	0
Mocatta, F. D., Esq., F.R.G.S.	1	1	0
Molyneux, Comm. W. H.	1	1	0
Moore, Major A.	5	0	0
Moorsom, Rear-Admiral	1	1	0
Morant, George, Esq.	2	2	0
Morgan, Lieut. W. G. H., R.N., H.M.S. *President*	5	0	0
Mundy, Admiral Sir George, K C.B.	6	0	0
Murray, John, Esq., F.R.G.S.	10	10	0
Norcliffe, Colonel	1	1	0
Ommanney, Capt. E., R.N., F.R.G.S.	2	2	0
Ommaney, M. C., Esq.	1	1	0
Otway, Comm. Robert Jocelyn, R.N.	10	0	0
Otway, Sir George, Bart.	1	1	0
Outram, Sir Benjamin, C.B., F.R.S., F.R.G.S.	5	5	0
Ouvry, Frederick, Esq.	1	1	0
Overstone, Lord, M.A., Pres. S.S., F.R.G.S.	25	0	0
Owen, Mr. (*Phœnix*)	0	10	0
Paget, Capt. Lord Clarence, R.N.	5	0	0
Pakington, Right Hon. Sir J., Bart., M.P., F.R.G.S.	5	0	0
Pallinson, H. L., F.R.S.	5	5	0
Palmer, Mr. (*Phœnix*)	0	10	0
Parish, Sir Woodbine, K.C.H., F.R.G.S.	5	5	0
Parker, Capt. John	1	0	0
Parr, Thomas Clements, Esq.	2	2	0
Parry, Rear Admiral Sir Ed., D.C.L., F.R.S., F.R.G.S.	3	3	0
Pasley, Lieut.-Gen. Sir Chas., R.E., K.C.B., F.R.S, F.R.G.S.	5	0	0
Pasley, Capt. Sir Thos. S., Bart., R.N.	1	0	0
Peacocke, Col. Thos.	5	5	0
Pearce, Stephen, Esq.	2	0	0
Perigal, F., Esq.	1	1	0
Petrie, Samuel, Esq.	2	2	0

NAMES OF SUBSCRIBERS.

	£	s.	d.
Phillips, T. J., Esq.	1	1	0
Phœnix, H.M. Steamer, Ship's Crew	2	3	7
Pinhorn, William, Esq.	1	1	0
Playfair, Dr. Lyon, C.B., F.R.S.	2	2	0
Pole, Major	1	1	0
Ponza, Chas., Esq.	3	3	0
Portlock, Lieut.-Col. J. E., R.E., F.R.S., F.R.G.S.	1	1	0
Pottinger, Lieut.-Col. W.	1	1	0
Powles, J. D., Esq.	10	0	0
Prescott, Rear-Admiral Henry, C.B.	2	2	0
Probyn, Capt. Geo., R.N.	1	1	0
Pym, Major-Gen. Sir Henry	5	0	0
Radstock, Rear-Admiral Lord, C.B., F.R.G.S.	3	3	0
Ramsden, Frank, Esq.	1	0	0
Raper, Henry, Esq., R.N., F.R.G.S.	1	0	0
R. D.	1	0	0
Registrar-General of Seamen	1	1	0
Renouard, Rev. G. C., B.D., F.R.G.S.	1	0	0
Renwick, Mr. (*Phœnix*)	0	10	0
Rich, Comm. H., R.N.	1	1	0
Richards, Mr. (*Phœnix*)	1	0	0
Rodger, W., Esq.	2	2	0
Roget, Dr. P. M., F.R.S., F.R.G.S.	1	1	0
Roscow, Comm.	1	0	0
Rous, Rear-Admiral the Hon. H. J., F.R.G.S.	10	0	0
Royal Artillery (Non-commissioned Officers employed in Col. Sabine's office.) *One day's pay*	0	10	3
Russell, Lord John, M.P., F.R.S., F.R.G.S.	20	0	0
Russell, J. Watts, Esq., D.C.L., F.R.S., F.R.G.S.	3	3	0
Sabine, Col. Edward, R.A., F.R.S., F.R.G.S.	5	0	0
Sainsbury, J., Esq., "the Collector and Proprietor of the Napoleon Museum"	10	10	0
Ditto, towards the Family Fund	10	10	0
St. David's, Bishop of	5	0	0
Salomons, Mr. Alderman David, F.R.G.S.	25	0	0
Sandwith, Lieut.-Gen., F.R.G.S.	5	0	0
Saurin, Rear-Admiral Edward	2	0	0
Sewell, Fox, and Sewell, Messrs.	3	3	0
Simpkinson, Lady	1	1	0
Simpkinson, J. M., Esq.	5	0	0
Slade, Felix, Esq.	5	0	0
Smith, Newman, Esq.	2	0	0
Smith, E. Osborne, Esq., F.R.G.S.	1	1	0
Smyth, Rear-Admiral W. H., K.S.F., D.C.L., F.R.G.S.	1	0	0
Sotheby, Capt. E. S.	1	0	0
Spence, William, Esq., F.R.S.	5	0	0
Spencer, Rear-Admiral the Earl, K.G., C.B., F.R.G.S.	25	0	0
Stanton, George, Esq.	1	1	0
Stanton, Mr. (*Phœnix*)	2	0	0
Stanton, C. H., Esq.	0	10	0

NAMES OF SUBSCRIBERS.

	£	s.	d.
Staunton, Sir G. T., Bart.	5	0	0
Steele, Lieut.-Col. Thomas, F.R.G.S.	3	0	0
Stirling, Rear-Admiral Sir J.	1	0	0
Stokes, Capt. J. Lort, R.N., F.R.G.S.	1	1	0
Sykes, Vice-Admiral John	2	0	0
Tatham, Comm. Edward	1	1	0
Thomas, Vice-Admiral Richard	3	0	0
Thompson, Frederick, Esq.	2	2	0
Thompson, Capt. J. T., R.N.	1	1	0
Tolver, S., Esq.	1	1	0
Townshend, J. V. S., Esq.	1	1	0
Towry, G. E., Esq., F.R.G.S.	1	10	0
Tremenhere, Seymour, Esq.	1	1	0
Trevelyan, Sir W. C., Bart., M.A., F.R.G.S.	5	0	0
Tudor, Henry, Esq.	2	2	0
Tudor, Edward Owen, Esq., F.R.G.S.	2	2	0
Turner, Henry, Esq.	5	0	0
Urquhart, James, Esq.	2	2	0
Vaughan, Colonel Wright	1	1	0
Vernon, John, Esq.	25	0	0
" Vivent les Bellots "	2	0	0
Vyner, Robert, Esq.	5	0	0
Waite, J. C., Esq.	1	1	0
Waley, S. W., Esq.	1	1	0
Walker, Capt. Robertson	2	2	0
Walker, Capt. W. Harrison, H.C.S., F.R.G.S.	2	2	0
Washington, Capt. John, R.N., F.R.S., F.R.G.S.	2	2	0
Weld, C. R., Esq.	1	1	0
West, Admiral Sir John	5	0	0
Whiston, Rev. Robert	1	1	0
Wilkinson, Conrad, Esq.	2	2	0
Wilkinson, Horace, Esq.	2	2	0
Wilkinson, Norman, Esq.	1	1	0
Wilkinson, William, Esq.	1	1	0
Williams, Rev. D., D.C.L., F.R.G.S.	1	1	0
Willich, Charles M., Esq., F.R.G.S.	1	1	0
Winslow, Dr. Forbes, D.C.L.	3	0	0
Wood, Sir W. P.	5	0	0
Wynniatt, Lieut. (*Phœnix*)	2	0	0
Young, Thos., Esq., F.G.S.	5	0	0
Younghusband, Capt., R.A., F.R.S.	1	0	0

NAMES OF SUBSCRIBERS.

SECOND LIST.

(Additional Names of Subscribers received at the Apartments of the Royal Geographical Society since Nov. 21, up to Dec. 24.)

	£	s.	d.
A Lady and Captain Inglefield	0	19	10
Anstey, T. Chisholm, Esq.	1	0	0
Armstrong, Capt. W. H.	1	0	0
Ashwell, J., Esq., F.R.G.S.	3	3	0
Back, Sir G., R.N., F.R.G.S.	3	0	0
Baker, Rev. Talbot	1	1	0
Baker, W., Esq.	2	2	0
Bolland, Lieut. J. B., R.N.	1	0	0
Barghtwen, John, Esq.	2	0	0
Baring, Right Hon. Sir F., Bart., F.R.S., F.R.G.S.	25	0	0
Barlow, Rev. John, M.A.	1	0	0
Barnard, Herbert, Esq.	2	0	0
Bazley, F., Esq.	1	0	0
Bird, Capt. E. J., R.N.	2	0	0
Bolton, Mr.	0	5	0
BOLTON, CORPORATION OF—			
Arrowsmith, P. R., Mayor	1	0	0
Barlow, R. S.	0	10	0
Black, Dr.	0	10	0
Crook, Henry	0	10	0
Gray, John	0	10	0
Gray, William, Esq.	0	10	0
Hargreaves, Charles	0	10	0
Knowles, James	0	10	0
Makant, William	0	10	0
Rushton, T. L.	0	10	0
Stones, John, Esq.	1	0	0
The C. C.	5	0	0
The Town Clerk	0	10	0
Booth, John, Esq.	2	0	0
Botfield, Beriah, Esq., F.R.G.S.	5	0	0
Bourne, Mr.	0	2	6
Bouverie, Comm. F. W. P.	2	0	0
British Lion, collected from the Working Men at the	0	13	0
Brodie, Sir B., Bart., F.R.S., F.R.G.S.	5	0	0
Broke, Capt. Sir Philip, Bart., R.N., F.R.G.S.	5	0	0
Buckley, Major-General	2	0	0
Bulford, Lieut. J., R.N.	0	10	0
Buller, Drs. J. and W.	1	1	0
Burton, Capt. G. G., R.N.	0	10	6
Butler, Lord Charles	0	10	0

NAMES OF SUBSCRIBERS.

	£	s.	d.
CHATHAM AND SHEERNESS, SQUADRON AT—			
Percy, Vice-Admiral the Hon. Jocelyn, C.B., Commander-in-Chief	5	0	0
Diamond, H.M.S.—			
Mitchell, Lieut. A.	0	10	0
Hope, Capt. Superintendent, Sheerness	1	0	0
Horatio, H.M.S.—			
Jenner, Comm. R.	0	10	0
Fegen, F. J., Esq., R.N.	0	10	6
Juno, H.M.S.—			
Bradshaw, Lieut. R.	0	5	0
Donnelly, S., Esq., Surgeon	0	5	0
Gregorie, Mr. G., Midshipman	0	5	0
Wood, Chas. B., Esq., Assistant-Surgeon	0	5	0
Poictiers, H.M.S.—			
Gahan, Charles, Esq., Master	0	5	0
Hobbs, William, Esq., Assistant-Surgeon	0	2	6
Jones, Dr. H. B.	0	5	0
Morrell, Comm. A.	0	10	0
Penfold, Mr. Frederick	0	2	6
Whipple, Lieut. T. C.	0	10	0
Royal George, H.M.S.—			
Bond, Lieut. Thomas	0	5	0
Richards, Capt. Superintendent, C.B., Chatham	1	0	0
Price, Arthur, Esq., R.N.	1	1	0
Tower, Lieut. Arthur, R.N.	1	0	0
Robinson, Mr. S. G.	0	2	6
Coode, Lieut. F. P.	0	10	0
Comerford, Mr. W. T.	0	2	6
Freeland, Lieut. H.	0	5	6
Harvey, Lieut. H.	0	10	0
Hills, Mr. C. H. Q., Master	1	0	0
Meyell, Lieut. Fras.	0	10	0
Murray, Mr. D. S., Midshipman	0	2	6
Norman, Mr. A. M	0	5	0
Waterloo, H.M.S—			
Barlow, J. C. Esq., Master	0	5	0
Batty, Mr. E. C.	0	5	0
Bullock, Lieut. H.	0	10	0
Christopher, Lieut. F. B.	0	10	0
Cook, J. H., Esq.	0	5	0
Dickens, Lieut. S. T.	0	10	0
Ives, Mr. W. H.	0	3	6
Marsh, Comm. J. B.	0	16	6
Ricketts, Lieut. S. H.	0	10	0
Ring, Dr. T. E.	0	10	0
Stopford, Capt. the Hon. Montague	1	0	0
Winnecott, Mr. S.	0	4	0
Chesney, Major-Gen., R.A., F.R.G.S.	1	0	0
Clark, Sir J., Bart., M.D., F.R.G.S.	5	0	0

NAMES OF SUBSCRIBERS.

	£	s.	d.
Connop, Capt. H.	5	0	0
Cotton, W. S., Esq.	1	1	0
Cunliffe, R., Esq., jun.	1	1	0
D'Aguilar, Lieut.-Gen. Sir G.	5	0	0
Damer, Col. the Hon. D.	2	0	0
Deake, Capt. J.	1	1	0
De Mauley, Lord, F.R.G.S.	5	0	0
Denison, S. C., Esq.	1	0	0
Dixon, Charles, Esq.	5	0	0
Dobree, Comm. T. P., R.N.	1	0	0
Douglas, Comm. W. M. W., R.N., Coast-Guard, Kinsale	1	1	0
Du Cane, Lieut. F., R.E., F.R.G.S.	1	0	0
DUNDEE, Magistrates and Town Council of	5	5	0
Edwin, Comm. F., R.N.	1	1	0
Ellis, Comm. Wm., R.N.	1	0	0
Fairholme, W., Esq.	1	0	0
Fordati, James, Esq.	1	1	0
Forsyth, Comm. C. C., R.N., F.R.G.S.	1	1	0
Fifeshire	0	2	6
Fonblanque, R., Esq.	2	0	0
Friend to the Cause	1	1	0
Frith, J. Griffith, Esq., F.R.G.S.	1	1	0
Galton, Capt. Douglas, F.R.G.S.	1	1	0
Gevelad, Mr.	0	2	6
Gilbert, J. O., Esq., R.N.	2	0	0
Gladstone, Capt. J. V.	2	0	0
Glasgow, the Earl of	25	0	0
Gordon, G. Huntley, Esq.	0	5	0
Grace, Rear-Admiral Percy	1	1	0
Grant, Capt. Sir R., R.N.	20	0	0
Greenough, G. B., Esq., F.R.G.S.	2	0	0
Griffiths, G. R., Esq., F.R.G.S.	2	2	0
Gunnell, Comm. E. H., R.N.	1	0	0
Gurney, Hudson, Esq.	5	0	0
Halkett, Rev. D. S., F.R.G.S.	5	0	0
Halkit, Major I. T. Douglas	1	1	0
Hall, R., Esq.	3	0	0
Hamilton, Wm. I., Esq., F.R.G.S.	1	1	0
Hammersley, Chas., Esq., F.R G.S.	2	0	0
Hamond, Capt. A. S., R.N.	1	0	0
Harnage, Edw. W., Esq.	2	2	0
Hayward, W., Esq., Dep. Comm.-Gen.	2	0	0
Henri, Lieut. Alphonso, R.N., and the Men of the Coast-Guard in the Killala District	1	2	6

NAMES OF SUBSCRIBERS.

	£	s.	d.
Henry, Graves, and Co.	1	1	0
Herries, Messrs.	10	10	0
Holland, Capt. F., R.N.	2	0	0
Holte, R. Orford, Esq.	2	0	0
Hosken, Comm. F., R.N.	0	10	0
Hubbard, G., Esq., F.R.G.S.	5	0	0
Hughes, Mr.	0	2	6
Hyndman, J. B., Esq.	5	0	0
Jennings, W., Esq.	1	0	0
Jerrold, Douglas, Esq.	1	0	0
Johnston, A. K., F.R.G.S.	1	1	0
Johnstone, Comm. F. E., R N.	1	0	0
Johnstone, H. J., Esq.	10	10	0
Keeling, E. H., Esq.	1	0	0
Kent, J., Esq., F.R.G.S., Assis. Com.-Gen. (1 day's half-pay)	0	9	6
King, Comm. Henry, R.N.	0	10	0
King, W., Esq.	2	2	0
Law, Hon. H. Spencer, F.R.G.S.	1	1	0
Lemon, Mark, Esq.	1	0	0
Lewis, Capt. G. C. D., late of the R.E.	1	1	0
Lightly and Simon	2	2	0
Lloyd's, the Members of	50	0	0
Lock, Comm. Cambell, R.N.	1	0	0
Lymington, Lieut. E. Powys, R.N., for the Monument	1	0	0
Lymington, Mrs. Story, for the Family	2	0	0
Macdonald, Lieut.-Col. John (through Messrs. M'Gregor)	1	0	0
M'Gee, W., Esq., M.D., R.N., Mayor of Belfast	2	2	0
M'Gregor, Capt. R. C.	1	1	0
Major, R. H., Esq., F.R.G.S.	1	1	0
Mayhew, Horace, Esq.	0	10	6
Micklethwaite, F., Esq.	1	1	0
Mills, C. Esq.	10	0	0
Moore, Mr., and the Men of the Coast-Guard in the Lochrus District	1	11	0
Morris, Huson, Esq.	5	0	0
Mountain, W. J., Esq., late of the Admiralty	1	1	0
Murray, Lady James	1	0	0
Northwick, Lord	10	0	0
Ommanney, Lady	1	0	0
Oswell, W. Cotton, Esq., F.R.G.S.	1	1	0
Peters, Mr.	0	1	0

NAMES OF SUBSCRIBERS.

	£	s.	d.
Peters, T. G., Esq.	2	0	0
Porter, Edward, Esq., F.R.G.S.	1	1	0

PRESS.

- Athenæum
- British Banner
- Daily News
- Examiner
- Globe
- Literary Gazette
- Morning Advertiser
- Morning Chronicle
- Morning Herald
- Morning Post
- Nautical Magazine
- Nautical Standard
- Navy List
- Sun
- Spectator
- Sunday Times
- Times
- Weekly Dispatch

} A certain number of Advertisements free.

	£	s.	d.
Purcell, Capt. E., R.N.	1	0	0
Raven, Comm. M.	1	1	0
Richardson, Sir J., C.B , F.R.S., F.R.G.S.	2	0	0
Rickardson, G. G., Esq.	1	1	0
Rodd, J. R., Esq., F.R.G.S.	1	0	0
Romilly, Edward, Esq.	5	0	0
Rosse, the Earl of, President Royal Society, F.R.G.S.	25	0	0
Rownley, Rev. Dr.	1	1	0
Royal London Yacht Club	5	0	0
Sach, Mr.	0	1	0
Sandeman, G. G., Esq., of 15, Hyde-park Gardens	1	1	0
Saunders, W. Wilson, Esq.	5	0	0
St. Asaph, the Bishop of, F.R.G.S.	2	2	0
St. Aubyn, Edward, Esq.	1	1	0
St. Leger, A. B., Esq., F.R.G.S.	5	0	0
Saurin, Rear-Admiral E.	2	0	0
Selwyn, C. J., Esq.	1	1	0

SHEERNESS District :—

	£	s.	d.
R. D. Aldrich, Comm., and the Officers and Men of the Coast-Guard	3	10	6
SHEFFIELD, Town Council of	20	0	0
Smith, H., Esq.	0	5	0
Smith, Wm., Esq., jun., of Bath	2	2	0
Spiller, J. H., Esq.	0	10	6
Stilwell, Messrs.	2	2	0
Streathfield, Lieut.-Col. C. O.	2	2	0
Strzelecki, Count P. E. de, C.B , F.R.G.S.	1	1	0
Stuart, Capt., J. C.	1	0	0

NAMES OF SUBSCRIBERS.

	£	s.	d.
Sulivan, Capt. B. S.	1	1	0
Swinburne, Capt. C. H., R.N., F.R.G.S.	5	0	0
Thornton, Samuel, Esq.	1	1	0
Townsend, Capt. J.	0	12	6
Trotter, Capt. H. D., R.N., F.R.G.S.	2	0	0
Tryon, Capt. R., R.N.	1	0	0
Tudor, Comm. John, R.N. (through Mr. A. Chippendale)	2	0	0
Tuelly, N. C., Esq.	1	1	0
Uncle Toby's Cabin	0	2	6
Vaux, W. S. W., Esq., F.R.G.S	1	1	0
Vincent, G. G., Esq.	1	0	0
Wall, Rev. M. S.	3	0	0
Ward, Geo., Esq., F.R.G.S.	5	0	0
Waterfield, Dr.	1	1	0
Webster, Capt.	1	0	0
Wedderburn, J., F.R.G.S.	1	0	0
Westhead, J. P. Brown, Esq., of Lea Castle	5	0	0
Weston, Alex. A., Esq., F.R.G.S.	1	0	0
White, W. E., Esq., F.R.G.S.	1	1	0
Wicklow, Earl of	5	5	0
Willes, James S., Esq.	5	5	0
Williams, Capt. B., F.R.G.S.	1	1	0
Willoughby, Comm. J. B., R.N.	1	0	0
Winthrop, Comm. H. E. S., R.N.	1	1	0
Woodhead and Co., Messrs., for—			
Blackwood, Capt. F. P., R.N., F.R.G.S.	1	1	0
Dorville, Comm. R.N.	1	1	0
Egerton, Capt. F. P., R.N.	1	1	0
Hall, Capt. W. H., R.N.	1	1	0
Hunte, Capt. Le, R.N.	1	1	0
Rice, Comm. E. B., R.N.	1	0	0
Pearse, Lieut., R.N.	1	0	0
Croft, Comm. H., R.N	0	10	0
Wolridge, Lieut. Chas., R.N.	1	1	0
Woodburn, William	5	0	0
Worms, S. B., Esq.	2	2	0
Young, Sir C. (Garter)	3	3	0
Yorke, Colonel Ph., F.R.S., Vice-President R.G.S.	3	3	0

NAMES OF SUBSCRIBERS.

THIRD LIST.

(UP TO THE PRESENT TIME.)

	£	s.	d.
Almach Baraugh, Esq.	1	1	0
Arctic Officer, Wife of an	0	10	0
A Surgeon	1	0	0
Aylmer, Vice-Admiral Lord	5	0	0
Ball, S., Esq.	3	0	0
Ball, —, Esq.	3	0	0
Bayfield, Capt. H. W., R.N.	2	0	0
Bedford, Capt. R. T., R.N. (sundry Subscriptions collected)	8	12	6
Beverley, Mr.	1	0	0
Bland, Lieut., A., R.N.	1	1	0
Blunt, R., Esq.	1	0	0
Brighton Birthday Club	1	7	0
British College of Health	5	0	0
Buckle, Admiral M.	3	3	0
Bruce, Rear-Admiral	2	0	0
Burney, Venerable Archdeacon	1	1	0
Burr, T., Esq.	1	1	0
Burrell, Sir C. M., Bart.	5	0	0
Burton, Capt. A. T., R.N.	0	10	6
Cator, Capt. J. B.	1	0	0
Cadogan, Vice-Admiral Earl	10	10	0
Castle School Society	2	0	0
Childers, Colonel M.	1	0	0
Christie, Samuel H., Esq., F.R.S.	2	2	0
Clarke, Lieut. E T., R.N.	0	2	6
Corry, Rear-Admiral A. L.	3	3	0
Compton, T. B., Esq.	5	0	0
Charlton, Dr.	1	0	0
Dantze, Capt. T A., R.N.	1	0	0
Derby, the Earl of	10	0	0
De Robick Hastings, S. T., R.N.	4	0	0
De Vries, J. N., Esq., Paymaster R.N.	1	1	0
Dicken, Lieut. H. P., R N.	0	6	0
Dickson, Capt. D. T., R.N.	0	10	0
Dondley, R., Esq.	1	0	0
Douglas, Major-General W.	3	3	0
D. S.	1	0	0
Dundee, Trinity House	3	3	0
Egerton, Capt. the Hon. F., R.N.	2	2	0
Exeter Corporation	5	5	0
Fellows, Sir C., Bart., Vice-President R.G.S.	5	5	0
Franklin, W., Esq.	3	3	0
Glyn, Sir R. C.	10	0	0
Goldsmid, A., Esq.	1	1	0

NAMES OF SUBSCRIBERS.

	£	s.	d.
Goodfellow, General	2	0	0
Gurney, Samuel, Esq.	50	0	0
Grenville, C., Esq.	5	0	0
Hancock, W., Esq.	1	0	0
Hamond, Admiral Sir G. E., Bart.	2	0	0
Hankey, Commander T. B.	1	0	0
Harvey, Henry, Esq., R.N.	0	10	0
Hawke, Lieut. B., R.N.	0	10	0
Higgs, Capt. W. H.	1	0	0
Hodgson, Capt. Ellis	1	0	0
Hood, Capt. G., R.N.	1	1	0
Howeled, J., Esq.	1	1	0
Ingledue, C. J., Esq.	0	10	0
Islington, A few Friends at	3	3	0
Jackson, Lieut. R. A., R.N.	0	5	0
Lang, J., Esq.	1	1	0
Little, Major, R. M.	1	1	0
Lyons, Rear-Admiral Sir E., Bart.	2	0	0
Lyster, Capt., R.N.	1	0	0
Mapleton, H., Esq., M.D., 3rd Dragoon Guards	1	0	0
Mello, J. A., Esq.	1	1	0
Monckton, General H.	3	3	0
Money, Rear-Admiral	0	10	0
Murchison, Kenneth, Esq.	5	0	0
Norton, Capt. T.	5	0	0
Oriental Bank, for Subscriptions received at Bombay	133	0	0
Parker, Lieut. Murray, R.N.	1	1	0
Parks, Lieut. A., R.N.	0	10	0
Parks, Lieut. A. E., R.N.	0	5	0
Parks, Lieut. A. J. A., R.N.	0	5	0
Paulson, Capt. I. T., R.N.	1	0	0
Paynter, Capt. J. A., R.N.	1	1	0
Bawdon, Lieut. Charles, R.N.	0	6	0
Reid, Comm.	0	10	0
Ricardo, —, Esq.	1	1	0
Richardson, James, Esq.	0	10	6
Robinson, C. M., Esq.	1	1	0
Rodgers, Geo., Esq.	1	1	0
Rogers, Wm., Esq.	1	1	0
Romilly, Sir J.	5	5	0
Ross, Rear-Admiral Sir J., H.C.B.	2	0	0
Rubige, Charles, Esq.	0	10	0
Sarjent, W., Esq.	5	0	0
Saxe, —, Esq.	5	0	0
Schomberg, Capt. C. F., R.N.	1	0	0
Sidney, Right Hon. J., Lord Mayor	5	5	0
Smart, Capt., R.N.	1	0	0

NAMES OF SUBSCRIBERS. 403

	£	s.	d.
Smith, John, Esq.	3	3	0
Snow, W. Parker, Esq.	1	0	0
Stavely, Thos., Esq.	2	2	0
Stuart, Comm. R.	1	0	0
Stoddart, Count R.	0	5	0
Synge, Lieut. R., R.N.	0	2	6
Stock Exchange	74	11	0
Taylor, John, Esq.	1	1	0
Thoresby, Lieut.-Col. C.	1	0	0
Towsy, Lieut. G. W.	2	2	0
Ward, H., Esq.	1	1	0
Warre, J. A., Esq., F.R.S., F.R.G.S.	5	0	0
W. D. L.	1	1	0
Wigram, O., Esq.	1	1	0
Wilkinson, Sir J. G., LL.D., F.R.S.	2	0	0
Wood, Capt. J., R.N.	0	10	6
Wood, Mrs. J.	0	10	6
Yale, Lieut.-Col. P.	1	1	0
York, Albion	2	2	0
Yule, G. U., Bengal Civil Service	2	10	0
Yule, Capt. H., do.	2	10	0

Manufactured by Amazon.ca
Bolton, ON